Design of Experiments for Process Improvement and Quality Assurance

by

Robert F. Brewer, P.E.

EMP

BOOKS

Engineering & Management Press
Institute of Industrial Engineers
Norcross, Georgia, USA
http://www.iienet.org

01 00 99 98 97 96   6 5 4 3 2 1

Library of Congress Cataloging-in-Publication Data
Brewer, Robert F., 1923-
    Design of experiments for process improvement and quality assurance / Robert F. Brewer
        p. cm. -- (Engineers in business series; 2)
    Includes bibliographical references and index.
    ISBN 0-89806-165-2
    1. Quality control -- Statistical methods. 2. Experimental design. 3. Analysis of variance. 4. Quality assurance. I. Title.
II. Series.
TS156.B74 1996   96-14873
658.5'62'015195--dc20   CIP

Director of Publications: Cliff Cary
Acquisitions Administrator: Eric E. Torrey
Book Editor: Forsyth Alexander

Illustrations by Kevin Edwards
Layout by Anne Leeds
Cover by Marty Benoit

ISBN 0-89806-165-2

Engineering & Management Press
25 Technology Park
Norcross, GA 30092-2988
USA

Table of Contents

Preface

Those of us who have used or are using a statistical approach to problem solving in industrial situations are aware of the advances made using this approach in other situations: agriculture, biology, chemistry, and especially in the social sciences. Advances occur from a need to evaluate more precisely increasingly complex problems. Although there is a clear demand for more individuals to be aware of statistical problem-solving methods, one need not be a statistician. Indeed, if presented clearly, the methods and approaches are not beyond most of those working in industrial situations as supervisors, technicians, or engineers.

The goal of this book is to introduce a fresh approach to quality improvement. The statistical foundations for design of experiments were provided by R. A. Fisher (1935), a British biologist, who introduced the concept and the statistical procedures known as the analysis of variance. These procedures provide an analysis of the total variation into meaningful component parts. What analysis of variance does not do is reveal anything about the shape of the distributions of the components and how these distributions relate to specifications. This is accomplished by graphical methods, which are crucial if one is to fully understand the experimental outcomes.

The purpose of a designed experiment is to manipulate and examine variables and determine their effects. It is a group of batch processes conducted under controlled conditions. Graphical techniques, such as control charts, box plots, and probability plots, provide a variety of information on a distribution and present statistical data that are easily understood by non-statisticians. This is especially useful when analyzing and presenting the results. Since the use of statistical terminology is often useful, the terms are defined when introduced.

This book is a guide for technicians, supervisors, engineers, and managers involved in the planning, implementation, and interpretation of industrial experiments, especially those who must work with quality professionals and statisticians. It is an introduction to using industrial experimentation for process improvement. The professional statistician often finds that management, engineers, and technicians have little understanding of the principles involved.

This book is written for those who must know the subject to work with others implementing the techniques. It assumes the reader is not familiar with mathematical statistics. However, since reference to statistical terms and their use are unavoidable, most of these terms are defined before they are used. The book utilizes graphical techniques where possible and shuns mathematical solutions. The techniques have been successfully tested in industrial environments. This is a practical book; it does not develop theory. It is the result of over forty years in the application of these methods in industrial environments.

To make the most of this guide, readers should first familiarize themselves with the overall data analysis techniques as outlined in the frontispiece. These are some of the graphical methods used in data analysis for quality improvement. A flow diagram for process improvement (see fig. 1.1) shows that introducing these techniques to the quality improvement process helps select those applicable to work. This is based on the need to know, the level of expertise, and degree of interest.

The statistical approach to problem solving and experimental analysis is responsible for extensive advances in the agricultural, biological, pharmaceutical, and social sciences. This has occurred primarily due to the educational levels of the participants. However, industrial education (engineers, technical schools, community colleges, and others) has been notably lacking in training students to use statistical methods for problem solving. The use of statistical methods has not been a part of the technical training of shop supervisors, technicians, engineers, and middle management. For modern manufacturing, methods of controlling and improving the quality of manufactured products have become essential.

The industrial experiment must be a form suitable for analysis. This can only be obtained by designing the experiment (in consultation with a statistician) with due regard to the (statistical) principles involved.

Acknowledgments

I am especially grateful to Dave Moen, Ron Asher, Earl Mustonen, and Dick Zwickle. I am but one of many in the field of Quality Control who are particularly grateful to Bonnie Small. Bonnie served as chairwoman of the Writing Committee for the Western Electric (AT&T) publication, the *Statistical Quality Control Handbook* (1956). We wrote up many of my applications as examples, particularly the Engineering Section. Special recognition goes to my Special Emphasis Team: Clyde Bunting, John Kunda, Bob Lukasiewics, Malcon Ducic, and Ilene Dimaggio. Also a heartfelt thanks to Pat Nahas and Dr. Richmond Johnson, experimental statisticians, both of whom spent many hours reviewing the text. Finally, I'd like to thank the staff at EMP—Eric Torrey, Forsyth Alexander, and Cliff Cary—for their editorial support, and designers Kevin Edwards and Anne Leeds for their illustrations and layout work.

Quality Improvement: By Data Analysis and Design of Experiments

Organization

MGR.

OPER. | MFG. | ENG. | PURCH.

COORDINATOR

ACTION TEAMS | QC CIRCLES

Problem Solving & Data Analysis

Probability Chart

Project Selection

Cause-and-Effect Diagram

Correlation

Box and Whisker Plot

Y .050 .040 .030 .020   #1 #2 #3 #4 #5

Pareto Analysis

Flow Diagram

$\bar{\bar{X}}$

R

Variable Control Chart

Evaluation

$$\sigma = \frac{SPEC - \bar{\bar{X}}}{3\sigma}$$

$$\sigma = \frac{\bar{R}}{d_2}$$

$$\sigma = \sqrt{\sigma_1^2 + \sigma_2^2 + \sigma_3^2 + \sigma_4^2 + \sigma_{res.}^2}$$

Formulas

Specifications

Zero Defects

Product Variability

Process Curves

Data Gathering

Defect Location	Tally	Total
A5	IIII	4
A6	III	3
C8	II	2
C9	IIII I	6
Total		15

Tally Sheet

Frequency 25 20 15 10 5 0
8.3 8.8 9.3 9.8 10.3 10.8 11.3 11.8 12.3 12.8
Gain in Db

Histogram

Process Improvement

		A1		A2	
		B1	B2	B3	B4
C1	01	3	4	0	6
	02	0	5	-6	4
C2	01	3	-4	4	-1
	02	5	-	3	2

Designed Experiment

Chapter 1

Introduction to Statistical Process Control

"It is clear, of course, that the relation of input-output is a consecutive one in time and involves a definite past-future order. What is perhaps not so clear is that the theory of the sensitive automata is a statistical one. We are scarcely ever interested in the performance of a communication-engineering machine for a single input. To function adequately, it must give a satisfactory performance for a whole class of inputs, and this means a statistically satisfactory performance for the class of inputs it is statistically expected to receive."

-Norbert Wiener, *Cybernetics*

Data analysis and design of experiments involve, among other things, problem-solving. This book has been organized around problem-solving for improving manufacturing processes. Figure 1.1 shows a *flow diagram for process improvement*. It shows the steps involved in solving problems and improving processes using statistical techniques. The sections in this book will follow this diagram.

Quality control for products manufactured in the United States has traditionally been done by the inspection system. Customer demands for a product or service manufactured by a process comprise this system (fig. 1.2). A series of steps or procedural inputs, including materials, machines, methods, and people make up the process. Once a product

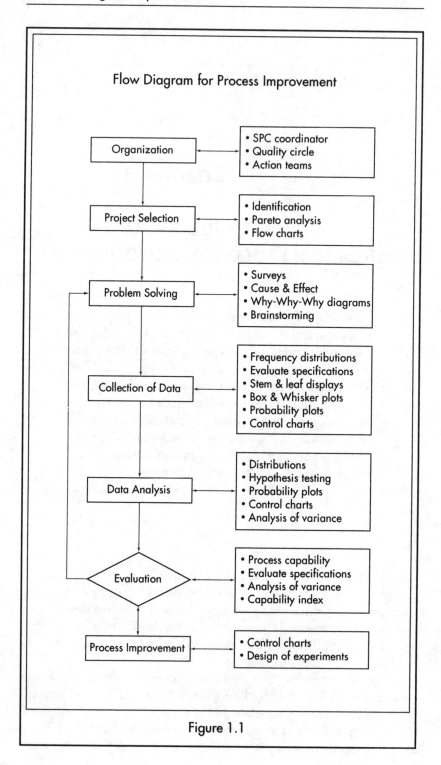

Flow Diagram for Process Improvement

| Organization | → | • SPC coordinator
• Quality circle
• Action teams |

| Project Selection | → | • Identification
• Pareto analysis
• Flow charts |

| Problem Solving | → | • Surveys
• Cause & Effect
• Why-Why-Why diagrams
• Brainstorming |

| Collection of Data | → | • Frequency distributions
• Evaluate specifications
• Stem & leaf displays
• Box & Whisker plots
• Probability plots
• Control charts |

| Data Analysis | → | • Distributions
• Hypothesis testing
• Probability plots
• Control charts
• Analysis of variance |

| Evaluation | → | • Process capability
• Evaluate specifications
• Analysis of variance
• Capability index |

| Process Improvement | → | • Control charts
• Design of experiments |

Figure 1.1

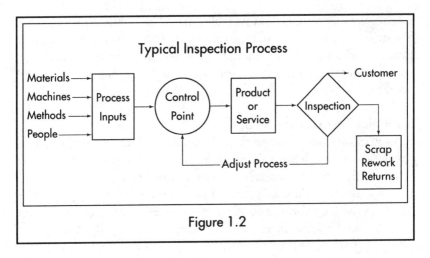

Figure 1.2

or service is produced, it goes to an inspection department that determines whether to ship the product or whether to scrap, rework, or dispose of the defects. The department may even determine if the defective part is usable.

A number of problems exist with the inspection system. Even with 100 percent inspection, defects are missed and sent on to the customer. If defects are discovered, the cost of correcting the defects has to be carried by the good product. Thus, either the customer absorbs it by paying a higher retail price or the company absorbs it internally. In either case, the inspection system is costly. There are also immeasurable expenses that occur when a customer receives a defective product.

The *Statistical Process Control* (SPC) system (fig. 1.3) consists of the same basic elements (customer, product or service, and process inputs).

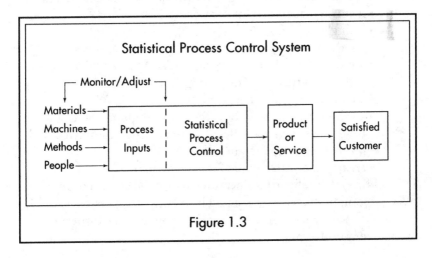

Figure 1.3

The *inspection process* is, however, eliminated or significantly changed. Statistical techniques monitor and adjust the processes at its source. Process variation is thus reduced by better raw material uniformity, better maintained machines, improved methods, and more careful operators.

Statistical process control

SPC involves the use of statistical signals to determine whether to change a process or leave it alone. There are several important principles:

- the process, product, or service must be measured;
- the data must be gathered as close to the source as possible;
- the monitoring and adjustments must be done by the person who knows the most about them, preferably the operator or machine setter;
- the data must be analyzed using statistical techniques such as control charts, probability plots, designed experiments, and analysis of variances; and
- action is taken before a bad product is made.

The term process, when used in SPC and Design of Experiments (DOE), means a system of operations that work together for a common purpose. While the processes usually referred to are often manufacturing processes, SPC and DOE are also applicable to service industries. They may or may not be controlled from a central point. However, all parts of the process should contribute to the production of optimum conditions (or outputs) for a given set of inputs, with respect to some statistical measure of effectiveness. For example, in a manufacturing plant, a series of machines makes and assembles a product. Since the process is a system of operations, it may be simple or complex. For example, it may be:

- a single machine, fixture, or operation;
- a single operator, clerk, or test;
- an assembly operation;
- a group of machines or spindles turning out similar pieces;
- a chemical treatment or plating system;
- a bank processing checks;
- a store receiving, storing, selling, and shipping merchandise;
- a combination of machines, operators, materials, and methods for making a product; or
- an airplane with a guided missile system.

A process may thus refer to the operation of a single cause, or it may refer to a very complicated cause system. This is why it is possible to apply statistical concepts to all types of operating, engineering, inspection, or management problems.

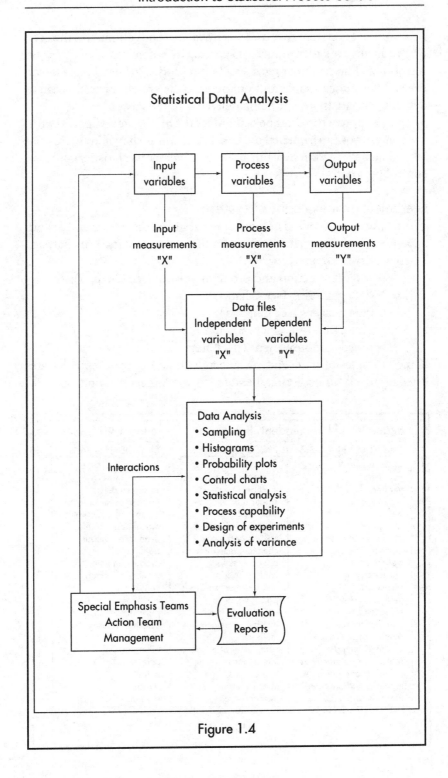

Figure 1.4

A process system that uses statistical methods eliminates the cost of producing a bad product, inspecting for defects, scrapping, or reworking. The customer receives a better product or service at a lower price. The immeasurable and unknown costs are eliminated because defective products are not being delivered to the customer.

The approach in this book combines the SPC system (fig. 1.3) with the concepts of experimental data analysis. This is shown in figure 1.4.

The next section explains how to implement a statistical process control program.

Statistical process control system

The measurements of the independent variables, process variables, and dependent variables make up the data files used for statistical analysis. These statistics are used for:

- project selection and problem-solving (see Chap. 3);
- data gathering (see Chap. 4);
- data analysis (see Chap. 5);
- evaluation (see Chap. 6); and
- process improvement (see Chap. 7).

Some examples of independent, dependent, and process variables are listed in table 1.1. This list may be added to from other operations, as well.

Examples of Independent, Process and Dependent Variables		
INDEPENDENT VARIABLES	PROCESS VARIABLES	DEPENDENT VARIABLES
materials	flaws in materials	number of defects
suppliers	thickness of materials	number of parts
vendors	mixtures	unknown defects
types of solders	solder flux	number of loose leads
welds	electrodes	number of poor welds
machines	machine maintenance	machine downtime
machine settings	speeds and feeds	production output
cleaning treatments	chemical activation	adherence
operators	operator skills	production time
chemicals	contamination	number of rejects
inspectors	measurement methods	yields
methods	part location	variations from specs
room temperature	assemble operation	parts not fitting
material source	number of defects	customer returns
number of operators	methods	costs
customer billings	complexity of forms	billing errors
materials and machines	interactions	defects
materials and machines	interactions	output
machine and operators	interactions	rejects
methods and machines	interactions	savings

Table 1.1

The most important measurements (or statistics) used in SPC are:
- the *average* ($\overline{X}$);
- the *range* (R);
- the *standard deviation* (s), and
- the *variance* (s^2).

Measurements on the input and process variables are introduced in the data file as independent variables (X). The output variables then become the dependent variables (Y).

The process model assumes that $\hat{Y}$ is a dependent variable, the value of which will change with any change in the independent variables. This is expressed in mathematical terms as $\hat{Y} = \text{ß}_0 + \text{ß}_1 X$ for populations, where:

- $\hat{Y}$ = the estimated dependent variable;
- ß_1 = the combined relationship between the dependent response variable $\hat{Y}$ and the independent variable X;
- X = the independent variable; and
- ß_0 = the parameter that indicates the average value of Y when X = 0.

Figure 1.5 shows the relationship, or *correlation*, between the dependent variables ($\hat{Y}$) and the independent variables (X).

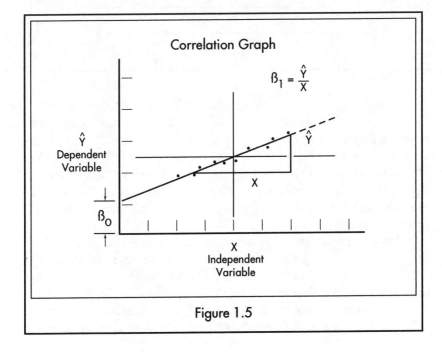

Figure 1.5

The scientific basis for designed experiments

Around 1840, Carl Fredrich Gauss laid the basis for statistics and the theory of probability–a result of his observations and measurements on the exact position, area, shape, and size of the earth. It came about through the thousands of sightings he had to make in his survey. Each measurement was made several times, often with different results.

By plotting the results of the readings, Gauss found that the graph always resulted in a bell-shaped curve (fig. 1.6), which today is often called the Gaussian distribution, or the normal, curve. He found that these plots followed a well-defined pattern: half of the errors were to the left of the center and half to the right.

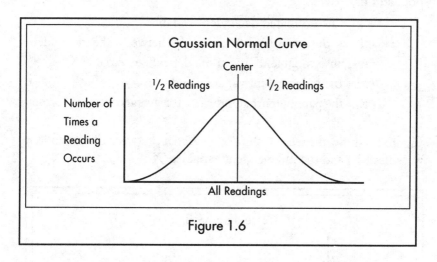

Figure 1.6

According to Gauss, these "accidental" errors were not chance. He felt that they must occur with a frequency that could be mathematically estimated in advance.

The Gaussian law of the normal distribution was later found to occur throughout nature. The observations obeyed the same law of large numbers. It is one of the most basic of all natural laws and includes the diameter of trees, the heights of people, the measure of industrial products, and so forth.

Weiner Heisenberg, an atomic physicist, won the 1932 Nobel Prize for Physics for formulating quantum mechanics in matrices. In what is popularly known as the "uncertainty principle" in quantum theory, he asserted that it is inherently impossible to determine the position and the momentum of a particle simultaneously and with unlimited accuracy. This principle is of great significance on the atomic scale. Its premise–that the location of a particle is intrinsically imprecise–forms the theoretical foundation of modern physics. This work led to a more precise theory about the structure of atoms and the effects of time on measurements (Andrews and Cokes, 1965). It is not possible to describe precision measurements without referring to time. The outcome of precision measurements has some error attached.

Heisenberg demonstrated that this uncertainty principle was applicable at the atomic level. A small disturbance that creates an uncertainty is a part of all measurements. This uncertainty in measurements is not meaningful for large bodies; if it were, it would invalidate the laws of mechanics. And so it must coexist with physical laws.

Jacques Monod, a Nobel Prize winning French biologist, in "Essay on the Natural Philosophy of Modern Biology," called the coexistence of probability and the laws of physics "chance and necessity" (Monod, 1978). He demonstrated that chances for mutations (a sudden change) occur at the genetic level. These chance variations are responsible for the broader framework of differences in populations, not in isolated individuals. This approach helps explain why the Gaussian distribution is applicable to molecular atoms–the building blocks of all matter. This is illustrated in figure 1.7.

The pictures of the atom sketched in figure 1.7 are models. Some regard an atom as a nucleus with a positive charge of electricity surrounded by electrons circulating like planets around the sun–the *particle* model.

For other purposes, an atom may be seen as a nucleus surrounded by a system of generalized three-dimensional waves–the *wave* model.

Heisenberg saw the atom as composed of infinite, linear "virtual" vibrations with frequencies–a matrix of interactions from finite to the infinite–the *matrix* model.

In the new quantum theory, either the wave-or-matrix form gave a perfect mathematical description of atomic phenomena. This, however, failed to illuminate the physical picture. The answer to this question was given by Heisenberg as *the uncertainty principle.*

The uncertainty principle states that the position and velocity of an object cannot be measured exactly, at the same time, even in theory. The very concept of exact position and exact velocity together has no meaning in nature.

Experience has shown that this normal distribution can be thought of as a composite mass of fluctuations. These fluctuations are confined within well-defined limits; therefore, it is possible to make use of the these limits. The limits are derived from the shape of the distribution.

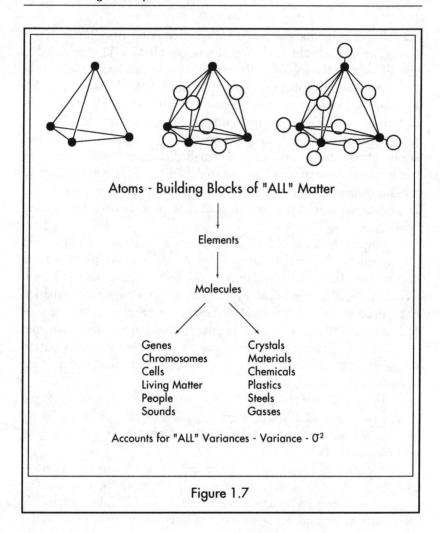

Atoms - Building Blocks of "ALL" Matter

↓

Elements

↓

Molecules

Genes   Crystals
Chromosomes   Materials
Cells   Chemicals
Living Matter Plastics
People   Steels
Sounds   Gasses

Accounts for "ALL" Variances - Variance - σ^2

Figure 1.7

The properties of the normal distribution have been thoroughly investigated. To make use of this distribution, it is illustrated in a tabulated table in Appendix A. Its use will be explained in detail in Chapter 5.

Translated into industrial terms, *normal distribution* can be thought of in the following manner: whenever observations or measurements are obtained from a given process, those measurements will not in general be identical with each other. If nothing disturbs the process the measurements will be held within definite limits. In their entirety, many of these measurements will tend to form a predictable distribution.

This can be generalized in the following terms:
- all things vary;
- individual things are unpredictable; and
- groups of things from a uniform system of causes tend to be predictable.

Process improvement: the Japanese contribution

Before World War II, the Japanese had a reputation for making inferior goods that only resembled those of better quality. The words "shoddy" and "junk" were equivalent to "made in Japan."

A normal distribution of Japanese manufactured goods of this period would look like that in figure 1.8.

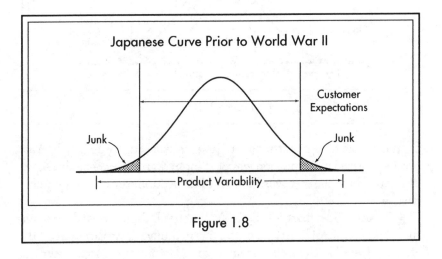

Figure 1.8

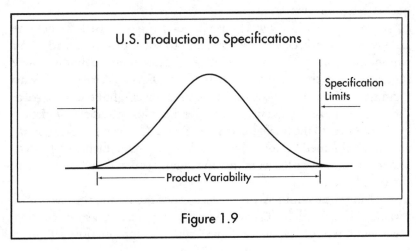

Figure 1.9

In the United States the emphasis was on producing a product to specifications. A normal distribution compared to specification limits on goods manufactured in the United States would look like that in figure 1.9.

The U.S. system of producing to specifications resulted in process variability, whereby a small percentage of products exceeded specifications (fig. 1.10). This resulted in a few products being identified

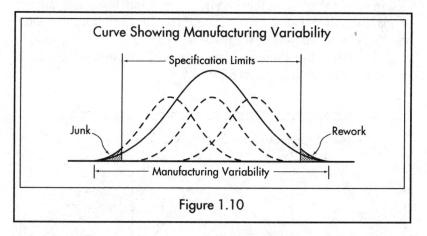

Figure 1.10

by the inspection department as junk, reworked, or rejected. Any product missed by inspection was paid for by warranties. The final cost of the product had to cover the cost of inspection, rework and junked parts.

Walter Shewhart of Bell Telephone Laboratories developed the $\overline{X}$ and R control charts to control this manufacturing variability. His efforts resulted in the Bell System earning a reputation for providing a quality telephone system.

During World War II, personnel from the Bell Telephone Laboratories worked with the Japanese to solve problems with their communication system as part of rebuilding their economy. They suggested that the Japanese obtain the services of Dr. W. Edward Deming to show them how statistical methods might be used to solve their problems basic to quality control. Dr. Deming introduced statistical quality control to the Japanese. He told them, "You can produce quality and here's the method for doing it. Bring your manufacturing process under control . . . with ever increasing quality you will capture markets the world over."

Using the techniques of designed experiments and the analysis of variance, Dr. Genichi Taguchi went a step further. By exceeding specifications, the Japanese found that the production variables

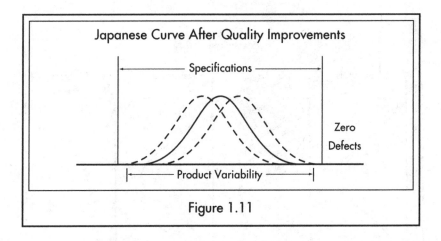

Figure 1.11

(fig. 1.11) did not result in junk or rework, and that they also eliminated warranty expenses. With these reductions in manufacturing costs, they were able to sell a superior product at less cost. They have now captured world markets just as Deming had predicted they would.

Process improvement with statistical process control (SPC)

SPC is a method for improving quality and productivity by not only doing it right the first time, but by also keeping the process under control. SPC uses a group of techniques that makes this improvement. As a result, the producer gets a better product that costs less to make, and the customer gets a better product that performs as intended for a longer period of time.

Properly understood and pursued, the goal of zero defects is vital for manufacturers. To meet this goal, it is important to have a full understanding of what the customer wants and how the product is to be used. This means well designed and understood specifications. This is the measure of effectiveness by which the process improvements are to be judged.

SPC is the application of the appropriate statistical techniques. Their implementation is made easier if an employee on the process line understands these techniques and works with his supervisors in their implementation. Unfortunately, the employee and his immediate supervisors usually do not have the background for understanding these fairly sophisticated techniques.

This book, however, uses easy-to-understand charts, graphs, and tables in the application of SPC technology. For example, process variability comes from many sources and each source of variability has its own unique effects, as well as the possibility of interacting with

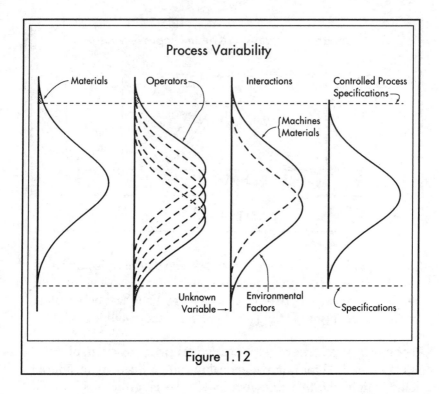

Figure 1.12

other variables. Figure 1.12 is an illustration of this point. An examination of this illustration shows that the part of the distribution outside the specification may be coming from different sources. Analysis of Variance, and DOE are used to determine what is causing this variation and what must be done to reduce or eliminate it from the process.

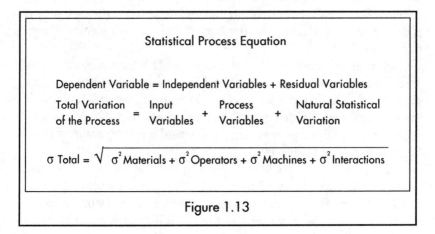

Figure 1.13

Figure 1.13 is an example of a statistical equation used to describe the effects of many variables on a total process. σ^2 is the statistic used to represent the process variation for each factor.

Conclusion

Process improvement comes about by identifying the contributing variables and finding ways to reduce them. Every job can be improved and statistical tools are universally applicable. In order to learn about the contributing variables, these variables have to be identified and measured. Some of the more important techniques for process improvement involve the following:

- Pareto Analysis;
- Cause-and-Effect Diagrams;
- Brainstorming;
- Why-Why-Why Diagrams;
- Flow Process Analysis;
- Probability Plots;
- X-Bar and R-Charts;
- Cusum Charts;
- DOE;
- Multiple Regression Analysis; and
- Analysis of Variance.

These are all part of process improvement technology. Most are discussed in the chapters that follow.

Chapter 2

Organizing and Implementing Industrial Experiments

"A mighty maze! But not without a plan."
- Alexander Pope, *Essay on Man*

The implementation of industrial experiments must be directed toward selecting the projects, solving problems, and collecting and analyzing data that will contribute to process improvement. This must be done with the full support, cooperation, and involvement of the plant manager because experimentation involves major changes. These changes are the responsibility of the plant manager and his staff. They must know what changes are required and see that these changes are carried out with a minimum of interference to plant production.

The implementation of industrial experiments is a management responsibility that must be controlled and methodically introduced into the manufacturing plant. Figure 2.1 illustrates a typical implementation plan.

Coordinator

The coordinator is the key person in the experimental program, providing the direction and leadership. He or she is responsible for:

- providing a continuity of purpose between all parts of the organization;
- organizing and scheduling the action and special emphasis teams and statistical consultants; and
- doing or securing the necessary training.

Implementation Plan

PHASE	ACTIVITY	RESPONSIBILITY	DESTRUCTIVE FORCES
START-UP	Funding Publicity Training	Top management Plant manager Coordinator	Inadequate funding Resistance to change Inadequate training Inability to learn problem solving skills

INITIAL
PROJECTS
	Process
	capability
	studies
		Coordinator
		SET team
			Disagreement on
			  selected projects
			Lack of knowledge of
			  statistical techniques

APPROVAL OF
INITIAL
PROJECTS
	Presentation
	of reports
		Plant manager
		& coordinator
			Resistance of staff and
			  middle management
			Limited knowledge of
			  statistical techniques

IMPLEMENTATION	Professional
	guidance
	Form action
	  teams
	Data
	collection
	Control
	charts
		Middle managers
		Coordinator
		
		Action teams
		
		Shop supervisor
		Foreman
			Resistance of shop
			  departments to
			  implementation
			Resistance to recom-
			  mended projects
			Inadequate training
			Failure to maintain
			  charts

EVALUATION	Interpretation
	of data
	Probability
	plots
	CPK reports
		Management
		Coordinator
		SET team
		Action teams
			Failure to interpret
			  data correctly
			Failure to identify
			  significant variables

IMPROVEMENT	Design of experiments Analysis of variance	SET teams Action teams Coordinator Management	Failure to run experiments

Figure 2.1

The coordinator should have the authority with the concurrence of the plant manager to stop or start production based on the results of an analysis.

Team organization

The preparations for industrial experiments must be directed toward both problem-solving and accomplishing the tasks as efficiently as possible. Organizing people for problem-solving requires a different type of organization than that required for getting the job done. Thus, process improvement and problem-solving must be structured to accomplish both courses of action (fig. 2.2). An action team is an effective way to do this. Figure 2.3 reflects the two basic types of team organizations: a traditional hierarchy team and a circle team. Whether a hierarchy team or a circle team is used for problem-solving, the members should be from various departments such as administration, purchasing, and sales, and should include engineers, statisticians, technicians, and shop supervisors.

The hierarchy team follows a chain of command under a designated leader. That leader assumes responsibility for the problem-solving process. He or she elicits responses from team members and encourages them to share relevant information. The leader tends to accept rather than evaluate contributions by clarifying and listing them. He or she periodically summarizes ideas to indicate progress.

Studies indicate that a hierarchy team is the best for solving easy problems and acting on them, particularly when the leader of the team has open lines of communication with all parts of the system. Ideas or solutions to a problem become accepted when they receive a certain amount of support. However, if a solution appears to have merit, then the solution may be accepted too hastily. Conversely, if a suggestion for improvement is made and the leader vetoes it, the matter is dropped.

A circle-type team depends on mutual trust and a willingness to share information. Solutions tend to be delayed. Disagreements are perceived as stimulants rather than threats. Circle members develop a problem-solving attitude. The circle lends itself very well to brainstorming (e.g., no one is permitted to criticize an idea, all ideas are encouraged, appraisal is delayed).

Among the studies available, one has shown that the circle team is best for solving difficult problems (Goode and Machol, 1957). When a suggestion was made in the circle team formed for the study, it was accepted, and the result was improvement. However, it should be noted that the members of that team felt they were disorganized and could have done better with a team leader.

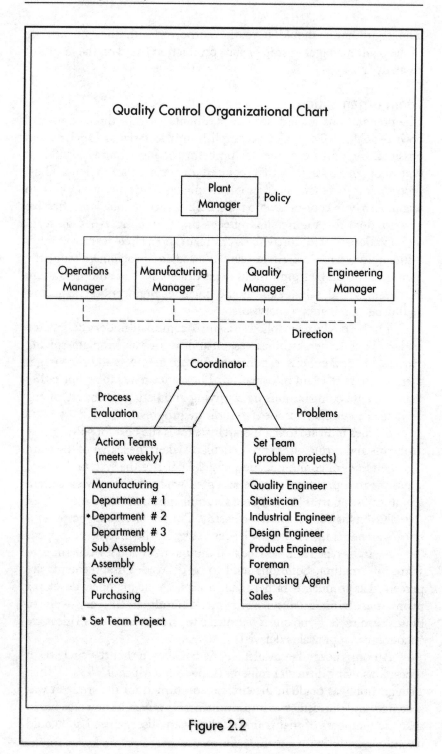

Figure 2.2

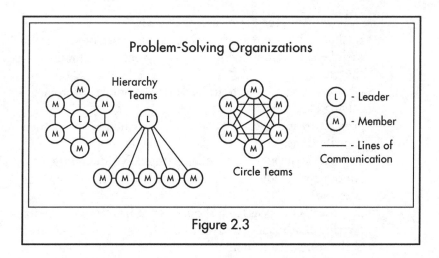

Figure 2.3

A *Special Emphasis Team* (SET) is a circle team made up of a small group of selected employees that meet to discuss special quality problems. They investigate causes and recommend corrective action and/or solutions. A special emphasis team makes an especially effectual action team.

SET members should be selected for their knowledge about the problems being investigated or for their expertise. These members could include, but are not limited to:

- supervisor of the production process;
- project engineer;
- machine operator;
- inspector;
- purchasing agent or buyer;
- designer of the machine, product, or process;
- maintenance man or toolmaker;
- union representative;
- sales representative; and
- those characterized by academic training in the scientific disciplines.

As can be seen from the above list, the SET should include members with a range of disciplines, skills, and experiences. All should bring a unique view point that aids in problem-solving and contributes to process and product improvement. This team is not like a normal organization. It represents a fundamental difference in approach because it brings together people who normally do not communicate with one another. Some are interested in procedural changes, while others are primarily interested in making equipment or material changes.

The following are the key elements of a SET:
- it should be a voluntary effort on the part of the group members. Everyone should participate but avoid dominating the group;
- group members should be trained in the techniques of problem-solving (see Chap. 3);
- projects are group oriented and the ideas are the sum of their contributions;
- creativity is encouraged. No idea is stupid. No idea is ridiculed. An outlandish idea may ignite a really hot idea in someone else; and
- the special emphasis team should not be held responsible for carrying out the suggested solutions. Ideas should be reported to the team's coordinator.

Action teams: organization

With the concurrence of the team coordinator and the plant manager, action teams are responsible for implementing process changes. They should meet on a regular, weekly basis to determine problem areas, assign projects to special emphasis teams, and be responsible for approving and implementing changes in the process. Each team should be headed by the person most knowledgeable of the process—usually the project engineer or shop supervisor. Figure 2.4 is an example of an

Example of an Action Team Organization

ACTION TEAM	DEPARTMENT HEAD	TEAM MEMBERS		
Purchasing	Buyer	Sales & Design	Engineering	Production
Department #1	Foreman	Union Sales	Industrial	Purchasing
Department #2	Foreman	Operator	Manufacturing	Sales
Department #3	Foreman	Plant Manager	Operator	Purchasing
Sub Assembly	Foreman	Industrial Eng.	Purchasing	Accounting
Assembly	Manager	Purchasing	Union Sales	Design
Service Dept.	Manager	Purchasing	Engineering	Quality
Inspection	Foreman	Sales & Design	Operator	Union

Figure 2.4

Project Sheet

SPC ACTION TEAM

Project Number _____

| Dept. _____ | Engr. _____ | Date _____ / _____ / 96 |

/--------------Part--------------\		Op. Drg.	/--------------Part--------------\	
Name	Number	Rev.	Rev.	Name	Number

Op. No.	Char.	S.P. Spec.	Insp. Method	Gage Number	Freq.

Problem/Reason for Selection

Sketch:

PAGE_____OF_____

Figure 2.5

action team organization. The team members should have some periphery knowledge of the product, its use, manufacturing, and sale.

Action teams should meet in an area where they will not be disturbed: a room designated specifically for these meetings. It should have a blackboard and be stocked with chart paper, probability paper, tablets, and pencils. If possible, a view graph should be made available.

To ensure an orderly approach to the activities of the action team, each project should be assigned a project number. Identify the part by number, machine number, operation number, characteristic, inspection methods gauge number, and its specific problem. Figure 2.5 is an example of a form that can be used for collecting this information.

The action team should also make out a weekly activity report that lists all of the active projects and the actions being taken. Figure 2.6 is an example of a weekly activity report. This report goes to the SPC Coordinator who forwards it to upper management.

An SPC action team project sheet (fig. 2.5) is used to record information taken during an action meeting. It is a record of the project that was discussed, what the problem was, and the reason for selection.

An SPC action team weekly activities report (fig. 2.6) is a report to management on the projects being worked. It includes status, results, benefits and future plans.

Administration of a process capability study

Process capability studies are used to determine if the process is operating normally. The coordinator should look for recognizable patterns that can be associated with the causes working in the process. If the disturbing causes are not immediately apparent, then control charts are placed on the production paths. That is, the data is split according to machines, shift operators, materials, or whatever is believed to be the controlling element. See figure 2.7.

Process control charts are used to control a process and determine when and how the process has changed (see Chap. 5). Control charts are also helpful in obtaining information about the process, and they track improvements in the process. Initially, the data should be collected into a histogram and plotted on probability paper (see Chap. 5 and fig. 2.9).

Probability plots are used to obtain information about the process distributions and batch processes. Initially, the data should be collected into a histogram from a frequency distribution of the data (see Chap. 4). A probability plot should then be constructed from the histogram data (see Chap. 5). Probability plots also are used to determine how the process distribution relates to specifications.

Project Sheet

SPC ACTION TEAM

Weekly Activities Report

Dept. _____ Date _____ / _____ / 96
Attendees _____
_____

Quality Status Report: No. of Parts/Operations _____
  No. of Charactoristics Added _____ Deleted _____
  No. of Charactoristics with CPK >1.3 _____ >1<1.3 _____
Status of Projects:

Proj. No.	Op. No.	Part No.	CPK	Remarks

Action to be Taken: (Status, Results, Benefits & Plans)
_____
_____
_____
_____
_____
_____
_____

New Projects:
_____
_____
_____
_____
_____

Figure 2.6

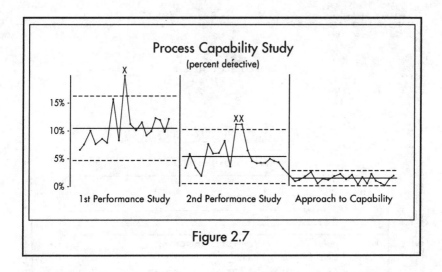

Figure 2.7

Introduction of process control charts

Action teams should be responsible for introducing the process control charts properly and making sure they work. The shop supervisor has a special responsibility, since he or she is the member of the team who works directly with shop employees.

It is best to prepare shop employees in advance for the introduction of control charts. The shop supervisor should hold a meeting approximately two weeks before introducing the process control charts. At this meeting, he or she should define the charts and discuss how they will benefit production and tasks. He or she should also explain how the charts will be made and how the samples will be taken, calculated, and plotted. He or she should emphasize the idea of making the product right the first time rather than having to repair it or sort it out by inspection.

It should also be stressed that all this will create a better product and reduce the cost of producing, which in turn will bring more business to the plant and ensure job security. Operators should be encouraged to ask questions about the charts and their possible effects on the plant and their jobs.

If possible, several meetings should be held to explain how control charts work, the meaning of the Xs on the charts, and what should be done when they appear on the charts.

The initial meetings should be followed by others after the charts have been introduced on the shop floor. The supervisor should point out the charts that show improvements, as well as any possible reasons for the processes that are in difficulty.

The success of shop charts is dependent on their being used to:
- help the operators and not to check up on them;
- obtain process information and control the quality;
- demonstrate unsuitable tools, fixtures, and machines and make it easier to correct the problem areas; and
- show the good work an operator may be doing.

Use of probability plots

Process variables produce all kinds of distributions that result in part of the process being outside the specification limits (fig. 2.8).

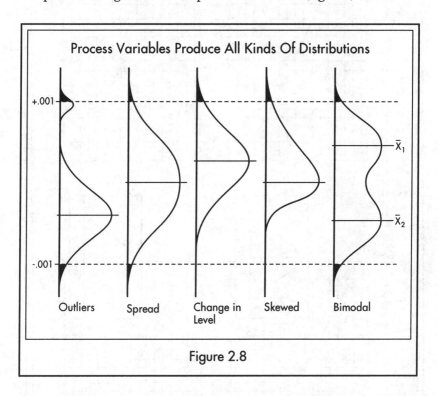

Figure 2.8

Probability plots that record process data on special graph paper provide a variety of information on distributions (fig. 2.9). A normal distribution plots as a straight line. A line, curve, or deviation from a straight line suggests possible problems such as a skewed distribution, a truncated distribution, a multimode distribution, or outliers.

A probability plot can provide a variety of statistical information about a distribution, such as average value, standard deviation, percentiles, and distribution limits. Most importantly, these plots relate the process distribution to specifications. Figure 2.10 illustrates these parameters.

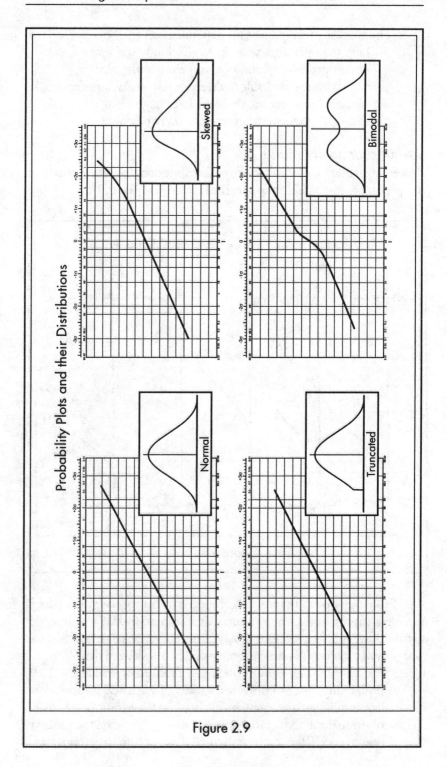

Figure 2.9

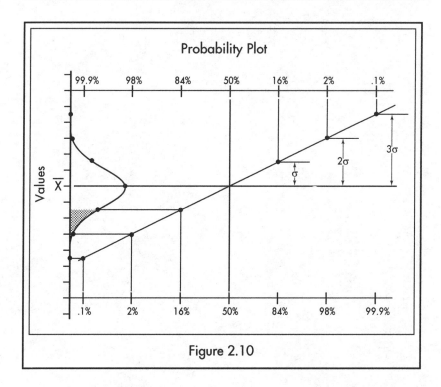

Figure 2.10

Probability plots, when used with control charts in a process capability study (see fig. 2.7), provide important clues to process improvement. Special emphasis teams should be able to present and interpret probability plots to action teams for their consideration.

Probability plots are covered in greater detail in Chapter 5.

Conclusion

Three of the four main techniques for SPC have been discussed above. The fourth is DOE. The SET team should be given the responsibility of designing experiments and overseeing the experiments. DOE and the interpretation of the results by analysis of variance provide the basis of new knowledge about a process. Therefore, the team must be trained in the principles and techniques of design of experiments and analysis of variance. The results of the experiments must be carefully observed and analyzed by a specific method.

Designed experiments and their analysis are covered in detail in Chapter 7.

Chapter 3

Project Selection and Problem-Solving

"Quality–What kind is it?"

-Plato

"Quality is never an accident, it is always the result of high intention, sincere effort, intelligent direction and skillful execution; it represents the wise choice of many alternatives."

-William A. Foster

"The problems . . . while elementary, are nevertheless complicated enough to require real skill in concentration and visualization, if one is to solve them in his head."

-Lewis Carroll, *Pillow Problems*

The best route to quality improvement is statistical problem-solving. Process variables produce all kinds of distributions that result in defective products. The initial step involves monitoring the process with statistical methods that identify the operating distribution. After the distribution of concern is identified, then the cause of the defective product must be determined.

The root definition of the word quality, which comes from Plato, is *what kind is it?* It implies that a customer is asking the question. Thus, the customer is involved in defining quality. A customer is anyone to whom an individual provides service, information, support, or product.

Customers include people in production, inspection, purchasing, sales personnel, and government agencies. They are the users of the product or service.

Identification of the customer

Who are the customers and what are their needs? Before one can identify the customer requirements, the customer must be identified. This is a key factor in a process improvement program. Basic to quality analysis is the fact that everyone in an organization–every person, every department, every level–has a customer. When individuals have a clear understanding of who their customers are and what they want, then real "customer service" is possible. To improve quality, one must find out what the customer wants that he or she might not be getting and then–whenever possible–provide it. Employees that satisfy customer needs tend to be motivated, committed, and productive. Organizations that satisfy customer needs usually are more successful and profitable than those who do not.

There are a variety of techniques for conducting customer research. Frequently, a combination of them provides the most valid results.

A simple graphic device for identifying customer needs is the customer window grid, an example of which is shown in figure 3.1 (Arbor, 1987). It is based on quadrant analysis, a market research tool. The grid divides product features into four quadrants or groups:
  • the customer wants it and gets it;
  • the customer wants it and does not get it;
  • the customer does not want it and gets it; and
  • the customer does not want it and does not get it.
For example, if a new product is being designed, potential attributes or features of this product can be placed in the appropriate quadrant. Knowing in which quadrant a feature or attribute falls can guide decision-making processes. Energy should be devoted to those features a customer wants or needs but is not getting. If the customer wants it and is getting it, then there is no problem; the feature is meeting his or her needs. If the customer is getting a feature and does not want or need it, it may be possible to eliminate or alter the feature. If it is not delivered and is not important, there is little reason to be concerned.

The customer window grid provides a concrete, visual guide for discussions about quality and the value of a product's attributes. This analysis is applicable for day-to-day decisions about internal customers, as well as for major critical decisions related to external customers.

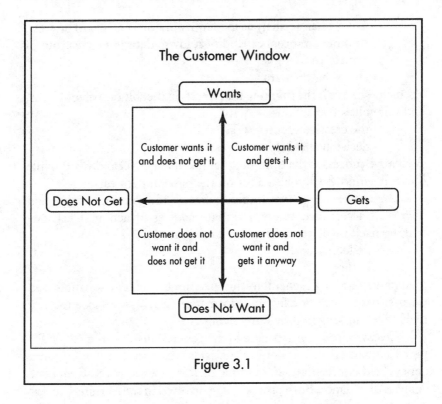

Figure 3.1

The customer window model should be used as a guide when collecting data. The most common collection methods are:
- interviews–listening to the customer;
- written questionnaires–a test of customer perceptions;
- focus groups–identifying group reactions; and
- indirect sources–sales representatives, purchasing agents, and others.

Problem-solving
Before searching for solutions, one must know what the problem is. A well-stated problem is half solved. What is known and what is unknown? The connection or relationship between the known and the unknown data must be found.

When defining a problem, these approaches are useful:
- targeting the area of interest (what has changed?);
- drawing a random sample of the area of interest;
- collecting data from the following: *flow process charts, operation sheets, process instructions, specifications,* and *test results;*

- using a Pareto analysis to identify important areas of interest;
- running a process capability study to determine areas out of control; and
- analyzing the data.

Problem-solving is the thinking that results in the solution of problems, and it involves two actions:

- the creative process and
- decision-making.

A creative process is the thinking that results in novel and worthwhile ideas. It cannot be accounted for on the basis of past learning.

Decision-making is the thinking that culminates in the choice and degree of acceptance among alternate courses of action. Decision-making itself involves two types of thinking:

- deductive and
- inductive.

Deductive thinking (general to the specific) is problem-solving based on past experience or existing information. Surveys are examples of deductive thinking in problem-solving.

Inductive thinking (specific to the general) involves the search for new elements. It means getting the facts first and then coming to a generalized conclusion. An example of this method is DOE and the analysis of variance. Both processes are involved in the learning process and problem-solving, as shown in figure 3.2.

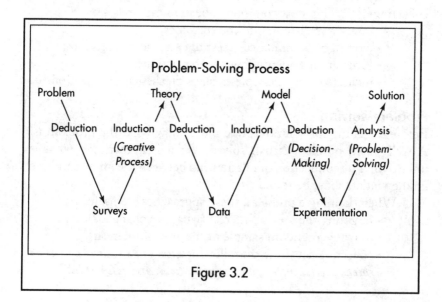

Figure 3.2

Among the range of techniques found to be helpful in promoting the problem-solving processes are the following:

- Pareto analysis–a graphical method for prioritizing projects;
- brainstorming–a method for creating new ideas;
- cause-and-effect diagrams–for identifying relationships;
- why-why-why diagrams–an aid in problem-solving;
- flow process analysis–a graphical representation of events;
- frequency histograms–a graphical arrangement of data;
- probability plots–a graphical display of distributions;
- scatter diagrams–a display of possible relationships;
- control charts–a graphical display of process data;
- capability study–a systematic study of a process;
- EVOP–running an experiment on a running process;
- DOE–deliberate experiments on a process; and
- Analysis of Variance–identification of process variables.

It is more important to find out the cause or causes of non-uniform behavior than to determine the percentage of defects from process inspection.

Experiments range from informal changes based on trial-and-error to carefully planned formal experiments. In order of increasing formality, these are:

- trial-and-error;
- running special lots;
- pilot runs;
- comparative experiments;
- designed experiments; and
- fractional experiments.

Trial-and-error experiments involve introducing a change to the process and then watching to see whether an effect shows up in the results.

When *running special lots*, identify the lots carefully with respect to the conditions under which they were chosen.

Pilot runs are experiments in which certain process elements are deliberately set up with the expectation of producing a certain effect. The results are then studied to see how close they come to what was anticipated.

Comparative experiments involve several factors.

Designed experiments are more complicated and involve a simple comparison of two methods.

Fractional factorial experiments are planned experiments that call for more factors arranged in even more complicated designs.

Study the initial steps of the process

In any process there is a progressive movement from one point to another on the way to completion. This development from a beginning to an intended end depends heavily on the first steps: that is, raw materials, purchased parts, and first operation. All have effects on each component of a continuing series of operations, acts, or development stages.

The materials and the first operations set the foundation of the process. If there is a problem in the first stage, it will inevitably show its effects in operations conducted at a later time. For example, if there is a defect in the raw material, it is most likely to appear during a machining or finishing operation. Therefore, SPC should always start at the beginning of a process and not be dependent on final inspection. This approach may also eliminate a later operation.

Surveys

The first phase of problem-solving is finding out what is already known about the problem by conducting a survey. Those with firsthand knowledge of the problems—engineers, operators, inspectors, supervisors, clerks, maintenance personnel, suppliers, and purchase agents—all can supply valuable information. These people can be instrumental parts of a problem-solving team formed to provide a cooperative atmosphere. The SPC coordinator and action team should initiate the data-gathering survey as follows:

1. *Analyze the problem.* Those involved must be specific when defining that problem to prevent false starts and "wheel-spinning." They should prepare a list of subjects to be investigated, noting the nature and extent of the information required. The problem should be carefully analyzed to determine what aspects need investigating. The problem may involve known physical and chemical laws (e.g., strength of materials, mechanics, chemical reactions, and others).

2. *Make a preliminary study of all unfamiliar aspects of the problem to be investigated.* For example, an engineering, chemical, or machinery handbook is a good source of information. In addition, it provides references to books, journals, and articles for greater details.

3. *Search for sources of information on all factors to be investigated.* The collection of books, periodicals, journals, indexes, and master reference catalogs in the company library is one source. However, a well-equipped technical library is often the best place to find sources. In it may be found reference books, abstracts, and index cards, including

the library index (alphabetic subject card file), the Engineering Societies Index (includes references for special technical fields), and periodicals (which also contain references and bibliographies).

4. *Make use of in-house specialists.* If the problem under investigation necessitates capabilities not available within the company, or an independent approach is desired, an outside consultant should be considered.

5. *Make use of known scientific principles.* Working with a mathematical or analytic model is an aid to problem-solving, not only because it identifies the type of data requiring research, but it also implies solutions. Examples of physical principles include:
    - strength of materials;
    - dimensional analysis;
    - transmission of heat;
    - machine stop practice;
    - properties of materials;
    - vibration;
    - dynamics;
    - friction;
    - feeds and speeds; and
    - stress analysis.

    Examples of chemical principles include:
    - chemical bonds;
    - crystal structures;
    - oxidation reactions;
    - nature of metals, alloys, and solutions;
    - chemistry of surfaces;
    - chemical equations;
    - electrochemistry; and
    - phase diagrams.

6. *Study, compare, and summarize the information gathered.* Is the information relevant? Has any essential data to be collected been omitted? The reliability of the data should be determined from practical considerations.

Pareto principle of analysis

A *Pareto analysis* is a bar graph that shows that a few items are usually responsible for most of the problems. It is a valuable tool for identifying the most important groups and ranking them. It is also a useful method of deciding problem priorities. It separates the *Vital Few* from the

Trivial Many. The Pareto analysis was named after V. Pareto, an Italian economist, sociologist, and engineer. Pareto demonstrated that the distribution of incomes tends to be similar despite governmental differences. When social classes were ordered, a few of the elite, about 20 percent, controlled 80 percent of the money. The Pareto principle has also been referred to as the 80-20 rule. During the study of manufacturing operations, it has been observed that 20 percent of the operations cause about 80 percent of the defects. A Pareto analysis then is a method of indicating which factors have the greatest impact on a problem. From it, decisions can be made as to where to direct efforts to decrease defects.

Figure 3.3 is an example of a Pareto chart for product defects. The heights of the vertical bars are proportional to the counts in each class. In this case, two defect types (dents and chips) are responsible for 76 percent of the defects.

A Pareto analysis has five basic steps:

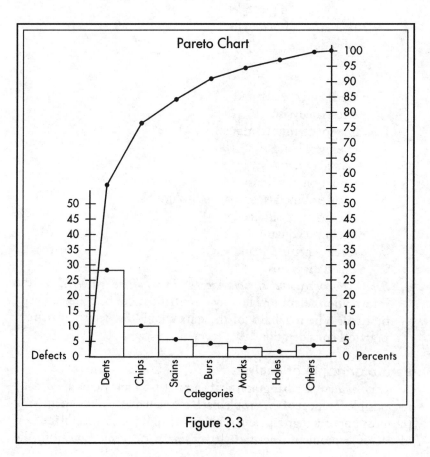

Figure 3.3

1. Data is collected by defect categories.

2. The categories are ranked by an attribute such as size or quantity of defects per category (see table 3.1).

Example Data

BY DEFECT	TYPE	DEFECTS	DEFECTS	PERCENTS
1. Most	Dents	28	28	56
2. -	Chips	10	38	76
3. -	Stains	4	42	84
4. -	Burrs	3	45	90
5. -	Marks	2	47	94
6. -	Holes	1	48	96
7. Least	Others	3	50	100

Table 3.1

3. A chart illustrating the nature of the relationship between the categories is prepared, usually shown on the horizontal axis (X), the abscissa, and the frequency that the event being measured occurs. Defects are shown on the vertical axis (Y), sometimes called the ordinate. Figure 3.4 illustrates the construction of this chart.

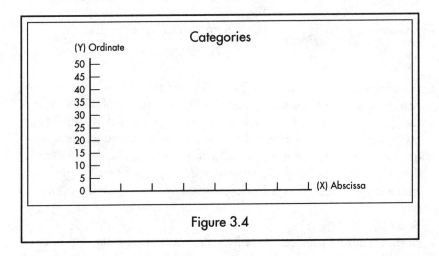

Figure 3.4

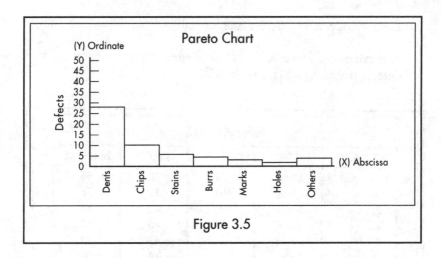

Figure 3.5

4. This step involves the ordering done in step 2. Category bars are used to represent the number of defects placed on the chart. The largest group is first, followed by the second largest, etc. If the data has only a few groups in the final categories, they may be combined into a final category labeled "other." See figure 3.5.

5. The cumulative percentages for the Pareto chart are determined. The number of defects is cumulated, the largest group first, followed by the second largest group, and so on.

Then, the percentages for each group are calculated. For example:

$$Cumulative\,Percents = \frac{Cumulative\,Defects}{Total\,Defects} = \frac{28}{50} = 56\%$$

A vertical axis on the right-hand side of the Pareto chart shows percentages. One hundred percent (100%) is at the top of the chart, and the rest of the scale is derived and labeled from that mark. At the midpoint of each category, a dot appears at the percentage level. The dots are connected (see fig. 3.3). In the example, the dent and chip defects are responsible for 76 percent of the defects. These are the items that need attention for quality improvement.

Brainstorming

Brainstorming generates new ideas and encourages creative thinking. Its main advantage is deferred judgment. All ideas, even unusual and impractical ones, are encouraged without evaluation or criticism.

Brainstorming sessions last from fifteen minutes to one hour and need no preparation other than general knowledge of the subject. Special emphasis teams allow brainstorming to be conducted in an interactive group environment. Before the session, the recorder gives the participants two to three minutes to jot down some of their initial ideas on the topic. Often a variation of one will turn out to be innovative or practicable.

Brainstorming imposes certain ground rules to prevent premature evaluation:

- no one is permitted to criticize an idea;
- silly, humorous, unusual, and impractical ideas are encouraged because they remove inhibitions; and
- no judgment should be made by the participants as it may inhibit productive thinking. If a member of the group does criticize an idea, he or she should be asked to make an alternate suggestion.

Problem-solving discussions are more productive if ideas are collected before they are appraised. The recorder notes the ideas as they are presented and writes them on a chalk board or easel for later appraisal. Colored markers are used to show relationships. Ideas are recorded as fast as they can be suggested.

Ideas generated during a brainstorming suggestion are evaluated and ranked at the end of the session.

Cause-and-effect diagram

Another aid in problem-solving is the construction of a *cause-and-effect diagram* (fig. 3.6), often referred to as a fishbone diagram. This easily generated diagram is helpful in selecting the important factors for analysis and their relationships. It is often used in the evaluation of ideas after a brainstorming session.

The backbone of the diagram may also be used to represent the main line of the production process and the skeleton for all things that may affect quality. This method involves going through the steps in the manufacturing process, one by one, to seek out the problem causes.

There are four steps involved in a cause-and-effect diagram:

- The major contributors are identified in categories of people, machines, materials, methods, and the environment. These main factors are the branches of the main arrow directed to the effect under consideration.
- The detailed factors considered to be causes are added onto the branches. Extended from these branches are more factors to be included as they come to mind.

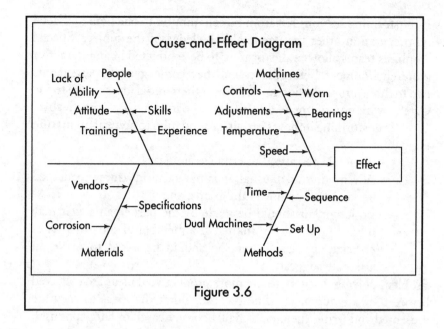

Figure 3.6

• All the items that may contribute to the effect are considered.
• The cause-and-effect diagram is evaluated by crossing out unlikely causes and circling the most likely causes.

Why-why-why diagrams

The *why-why-why diagram* (fig. 3.7) assists with brainstorming, cause and effect analysis, and problem-solving. A Pareto chart that has identified the most important contributors to the primary problem is a prerequisite. Four steps comprise a why-why-why diagram:

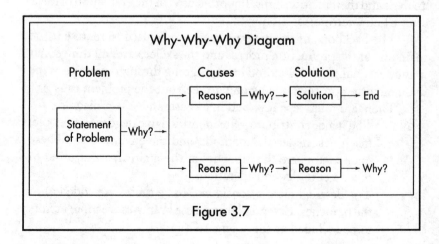

Figure 3.7

• a statement of the problem;
• statements of the causes (in answer to the question why). The causes may be separate or interrelated;
• each cause statement becomes a new problem statement, and again the question why is asked; and
• the chain of why questions continues to a solution.

Flow process analysis

A flow chart is a graphic representation of the events and information that occur in a series of actions or operations. The standard symbols used to represent these events are shown in figure 3.8.

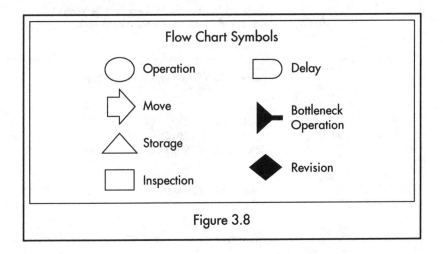

Figure 3.8

A graph allows the investigator to track and verify all operations in a process under analysis. When completed, a flow chart is filled with information not known before. The information collected at each operation should satisfy these questions: what? why? how? who? where? when? Figure 3.9 is an example of an operation flow process chart.

At operation 50 in figure 3.9, the part may be machined on any one of three spindles on the vertical milling machine. Any one of the three spindles could be a source of a problem occurring at a later operation. (It also offers the opportunity to compare statistically three equal manufacturing operations being done on three different machines.) Operation 70 has been identified as a bottleneck operation because of insufficient capacity.

A flow chart breaks down a process so that details can be visualized and the sequence of operations examined.

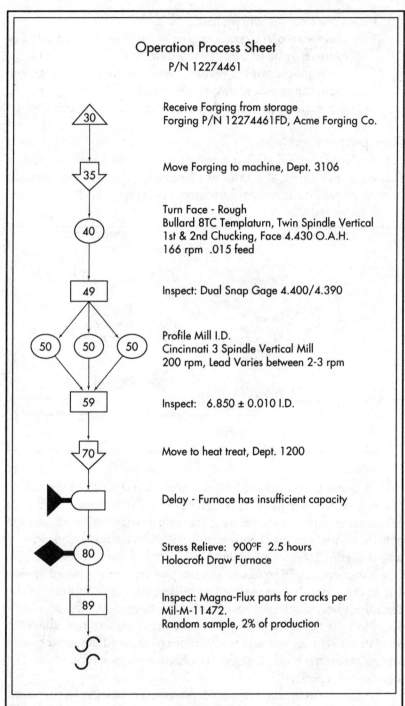

Operation Process Sheet
P/N 12274461

30 — Receive Forging from storage
Forging P/N 12274461FD, Acme Forging Co.

35 — Move Forging to machine, Dept. 3106

40 — Turn Face - Rough
Bullard 8TC Templaturn, Twin Spindle Vertical
1st & 2nd Chucking, Face 4.430 O.A.H.
166 rpm .015 feed

49 — Inspect: Dual Snap Gage 4.400/4.390

50 50 50 — Profile Mill I.D.
Cincinnati 3 Spindle Vertical Mill
200 rpm, Lead Varies between 2-3 rpm

59 — Inspect: 6.850 ± 0.010 I.D.

70 — Move to heat treat, Dept. 1200

Delay - Furnace has insufficient capacity

80 — Stress Relieve: 900°F 2.5 hours
Holocroft Draw Furnace

89 — Inspect: Magna-Flux parts for cracks per
Mil-M-11472.
Random sample, 2% of production

Figure 3.9

Crisis-generated projects

A project resulting from a crisis is often "self selecting." Therefore, it is given top priority and gets immediate attention. Problem-solving in a crisis is like collective bargaining: no one wants to lose and everyone wants to win. The worst possible outcome may be a bitter fight or stalemate.

A crisis interferes with the transfer of critical information often because it is withheld or overstated. When an experiment needs to be conducted, only a small change is made for fear of making the condition worse. First one thing is tried, then another; if neither works, something else is tried, and so forth.

Proceeding in small steps is inefficient and misses the possible effects of interactions of other variables. This is especially true when those who are conducting or interpreting the experiment do not fully understand that multiple experiments are the most efficient way to extract the most information. Actually, this is the right time for a designed experiment. It is the most efficient method of obtaining the maximum amount of usable information about a problem.

Breaking an impasse

When a team is unable to make headway in problem-solving, there are several methods that may provide an avenue to a breakthrough. The team can review the elements of the problem in rapid succession, several times, until a pattern emerges. Judgment should be suspended. Do not jump to conclusions. More evidence is required to overcome an incorrect hypothesis and establish a correct one.

Exploring the production environment is another method. Vary the time, space, and arrangement of the materials, working with flow process charts. The team could also attempt to regain flexibility by abandoning the approach and searching for a new basis for solving the problem. Open minds are essential; little time should be wasted on unsuccessful attempts.

The team should also effect a second solution, shifting the emphasis from a solution to the problem itself. All ideas should be critically evaluated with both the weak and strong points identified. If concrete representation is not working, try an abstract one and vice-versa (e.g., models, words, graphs, and numbers). Sometimes simply taking a break is helpful and trusting the subconscious to work on the problem is helpful. Finally, the problem could be discussed with someone outside the team. Aspects that might have been overlooked may arise from discussion. This is a powerful feedback mechanism because it forces a justification of procedures and results.

Conclusion

Every search for quality, particularly if problem-solving is used, needs a plan for investigation. Below is an example of an effective investigative methodology. This chapter has introduced the components; the rest will be discussed in subsequent chapters.

Plan for investigation

I. Statement of problem

A well-stated problem is half solved. Use the following to identify the problems, then summarize what has been gleaned from them: customer window, Pareto analysis, brainstorming, cause-and-effect diagrams, why-why-why diagrams, flow process analysis, and literature searches.

II. Collect the data

If it can be measured, then something will be known about it. To collect what is known about a process, make a frequency distribution, try a scatter diagram, and/or install a control chart.

III. Analyze the data

Examining the information collected often is only common sense reduced to calculation. The following can contribute to its understanding: histograms, box-and-whisker plots, control charts, and probability plots.

IV. Formulate a hypothesis

What can be concluded from the analysis? Analyze the variations with an F-ratio, t-test, probability plots, and control charts.

V. Test the hypothesis

Design an experiment and use known facts to prove the conclusions.

Chapter 4

Gathering Data

"Nature is full of infinite causes that never occurred in experience."

-Leonardo Da Vinci

"The purpose of data gathering is to obtain information. Good data (information) comes from asking good questions. We must learn to question, probe, investigate, and examine all aspects of the problem. Statistical data gathering is an active process. Listening is not enough."

-J. Stuart Hunter

The purpose of statistical process control is the improvement of a process. In that respect, SPC is concerned with the proper methods for collecting, organizing, summarizing, and analyzing data. Data is very important and appears in various forms.

The totality of items or units of material under consideration is called a population. A population, sometimes called a universe, consists of items that may be real or conceptual. While the terms population and universe can be and are frequently used interchangeably, a universe is sometimes used to mean a group of populations.

Some examples of populations are:
- the machine operators in a plant;
- a production lot of bearings;
- repeated measures on parts from a production process;
- parts produced on Mondays;
- repeated measurements from a particular CNC machine; and
- firings of rounds of ammunition from a given production lot.

In the first three examples, the components comprising the populations are material objects (operators, bearing, parts). In the fourth, they are from a period of time on a restricted day, and in the last, they are physical operations. The first two populations are clearly established and their members are determined by company records, while the third, fourth, and fifth are conceptually infinite. The last population would seem to be established, because firing is a destructive operation.

Populations may be large or small, but all should be well defined. The last two populations in the example are not completely defined until the measurements and firing procedures have been fully specified; however, definition eventually is achieved. The characteristics of interest must be the collective property of the targeted population.

It is often impossible or impractical to observe everything that is happening, particularly on a production line during a manufacturing process. Instead, a small, representative part of the population called a sample is examined. Conclusions regarding the process can be inferred from an analysis of the sample.

An industrial experimenter must collect data with only the slightest hindrance to normal production. Thus, it might be possible to obtain control of the independent variables, but not all the variables. Taking time and care on each production lot reduces the flow through the plant. It is possible to vary some of the independent variables in order to get data points over a wide range. Their effect on the dependent variable would then show up clearly; this, however, might cause the production of a large amount of scrap.

This is precisely why the data being collected must be subjected to statistical techniques and tests. Observations made on a process result in a body of data. Sometimes data is gathered to uncover a specific problem or to maintain control of the process. At other times, the nature of the information contained in a data set may not be obvious to those who requested or gathered the data. In either case, it is important to organize and summarize the data so that the results are more clearly understood. That depends on the nature of the data. Therefore, distinguishing between several types of data and methods of data collection is necessary.

Types of data

Data can be classified in two general ways. Data that is classified or counted is called attribute data. Measured data is called variable data (table 4.1).

Types of Data			
RANK ORDERED	ATTRIBUTE	SIMPLIFIED NUMERICAL	NUMERICAL
1	Excellent	6	22
2	Very Good	2	16
3	Good	-3	12
4	Poor	0	30
5	Bad	5	11

Table 4.1

Quality measurement by the attribute method consists of observing the presence (or absence) of a certain characteristic or attribute in the data group under consideration. Counting how many units do (or do not) possess the quality attribute, or noting how many events occur in the unit, group, or area are examples. The most common attribute measurement is the percentage of defects. Examples of attribute type data would be: good/bad, success/failure, go/no go, blue/green, greater than/less than, and percentage of non conformance.

If the type of data can assume only one value it is called a constant. A variable that can, theoretically, assume any value between two given values is called a *continuous variable*.

In general, measurements indicate continuous data while counting uncovers discrete data. Information from continuous data is called variables measurement, and this kind of data has the potential for a more penetrating analysis. Quality measurement by the variables method consists of measuring and listing the numerical size of each unit's quality feature in the group under consideration. This involves a reference to a continuous scale.

Historical data from plant records is useful for establishing past process or product trends. This helps identify areas for improvement and establishes a base line for measuring the improvements after changes are made. It is most frequently used in Pareto analysis, histograms, and attribute charts.

Data Collection Sheet
Tabular Format

PART NUMBER 2274825				PART NAME TORSION BAR		OPERATION (PROCESS) INDUCTION HEATING					
OPERATOR R. BREWER				MACHINE ACME INDICATOR		GAGE			SPECIFICATION LIMITS 98 ± .02		
						UNIT OF MEASURE % YIELD			ZERO EQUALS		

OBSER. NO.	MATERIAL SOURCE	Rc HARD	MACHINE NO.	CHAR. NO.	OPER. NO.	OPER.	SHIFT	TEMP °F	K.W.H.	DEFECT CLASS	YIELD
OLD PROCESS											
1	1	.54	1	60	122	JK	1	1600	.07	2	90
2	1	.53	2	60	122	SM	3	1600	.07	2	92
3	1	.52	1	60	122	JG	2	1600	.08	2	91
4	1	.55	2	60	122	SM	3	1600	.07	2	89
5	2	.54	1	60	122	MK	2	1600	.08	2	91
6	2	.57	2	60	122	SM	3	1600	.08	2	91
7	2	.60	1	60	122	JK	1	1600	.07	2	89
8	2	.62	2	60	122	SM	3	1600	.07	2	90
NEW PROCESS											
9	1	.55	1	65	122	JG	2	1650	.09	4	96
10	1	.57	2	65	122	SM	3	1651	.10	5	98
11	1	.56	1	65	122	MK	2	1650	.10	4	97
12	1	.58	2	65	122	SM	3	1649	.09	4	97
13	2	.61	1	65	122	JK	1	1650	.11	5	99
14	2	.60	2	65	122	SM	3	1651	.10	5	99
15	2	.62	1	65	122	JG	2	1651	.09	4	97
16	2	.63	2	65	122	SM	3	1650	.10	5	98

PLANT 7102   DEPT 3108   DATE 2-12-96   SUPERVISOR M. WILSON

Table 4.2

Current data is collected from present production at the time a product is being manufactured. It is used for identifying problem areas before "out-of-control" conditions are reached. Control charts are examples of current data.

Data collection

There are three common ways to collect data for SPC. They are:
- tabular formats or check sheets (table 4.2);
- hand-held data collectors (fig. 4.2); and
- fixed station data acquisition equipment (fig. 4.1).

All three are used on a factory floor, because each has certain advantages in certain circumstances.

Tabular formats

Data should be compiled in a form that will record all the known information about the process. A tabular format or check sheet is the best form. It should contain information necessary that associates the data with the product and process such as:
- identification number;
- identification name;
- operation;

Figure 4.1

Figure 4.2

• specification limits;
• machine name and number;
• gauge used;
• unit of measure;
• operator name; and
• date, department, and shift.

A drawing of the part with points of measurement clearly marked, as well as a process or operation sheet, should be attached to the data collection sheet.

Table 4.2 is an example of a tabular data collection sheet. Data was collected from a sample of eight parts on the old process and eight parts on the new process. The source and hardness of material, machine number, operation number, operator, operator shift, heat-treating temperature, power usage, and the defect class are all independent variables. The dependent variable is the percent yield in this example.

Columns are used to record groups of like data, and parallel columns should contain related data when applicable. Rows of data usually represent additional samples. The back of the sheet should be used for notes related to the data being collected. These notes are important; always remember that the purpose of data collection is to obtain as much information about the product or process as possible.

Control chart data forms are a special type of tabular format. Based on order of occurrence, these forms group data in rational subgroups of five. Usually, approximately twenty subgroups are collected. These forms also have space for the calculation of sums, averages, and ranges of the subgroups (see table 4.6).

Hand-held data collectors

Hand-held data collectors offer several advantages. They record data in a format that can be read by a computer. They can capture data directly from electronic gauges. Although there is an initial cost to be considered, the equipment becomes economical when used frequently. Comparing a hand-held data collector such as DataMyte® data collectors with the qualities of tabular data collection sheets (see table 4.2) can determine the suitability of each for various data recording tasks.

A feature that hand-held data collectors share with a tabular collection sheet is on-the-spot analysis. An advantage of a hand-held collector such as DataMyte is the numerical and graphical display of statistical data. In addition, data collectors check limits, alerting the operators with an audible beep when the data ranges out of specification. Formatted reports with summary information can be obtained by connecting the date collector with a printer. The data can also be transmitted to a computer for further analysis.

Figure 4.2 shows a DataMyte data collector and a video monitor being used at a machine tool.

Fixed station data acquisition equipment

Fixed station data acquisition equipment displays summary information, control charts, and other graphs on a monitor, giving the operator a means of identifying a problem when it occurs. Such equipment has a variety of applications, including statistical process control of dimensional characteristics and packaging weight measurements and control.

Fixed station equipment has several levels of sophistication. At the lowest level, it records data from a single source. Analysis is rudimentary and not compatible with data collected from other sources. In some cases, an operator must oversee data collection and either record it or transmit it to a computer for analysis.

Some types of fixed station equipment have outputs that are compatible with hand-held data collectors for periodic link-ups. An example of this is a weigh scale and an electronic gauge mounted on a fixture as shown in figure 4.1. A system such as this allows the operator to record data and also do a real-time analysis. In essence, it is both a data collection system and a quality control computer.

The more sophisticated types of fixed station equipment may be part of a test station or bays. Operation is automatic or semiautomatic, and a computer is either resident at the station or the equipment is directly linked to a computer. Their singleness of purpose makes them ideal for high volume tasks critical to process control.

Formatting data

Like every other science, statistical analysis and design of experiments have their own forms, words, and symbols with special meanings. Although at first this may seem confusing, but it is essential for the normal use of these methods. The following are the concepts used to handle data.

Significant numbers

Statistical analysis requires the use of numbers that have meaning. The value of the numbers carries weight in the analysis of the data. In any numerical computation, the possible or desirable degree of accuracy should be decided and the computation should then be arranged so that the necessary number of significant figures, and no more, is secured. Four place tables are the usual numbers found in statistics. Fractions are converted to decimals and rounded off. For example:

$$1\tfrac{1}{8} = 1.125 = 1.13 = 113$$

Significance tests are used to decide, with a predetermined risk of error, whether the population associated with a sample differs from the one that has been specified (see Chap. 5).

Coding data

Data may be easier to handle if it is coded before analysis and decoded after. Coding consists of adding, subtracting, multiplying, and dividing by some convenient number so that the database is a manageable size. Coding can also be used to eliminate decimals and negative numbers. The statistical analysis of data is not affected in any way. See table 4.3.

Coding attribute data

Attribute data and character type data may also be coded for numerical analysis. If at all possible, attribute data should be classified or weighted so that more information may be gleaned from it. Table 4.4 offers an example.

Example of Data Coding

(Original Data Minus Constant) x 10,000 = Coded Number)

(3.1929 - 3.1890) x 10,000 = 30

(3.1910 - 3.1890) x 10,000 = 20

(3.1892 - 3.1890) x 10,000 = 2

(3.1915 - 3.1890) x 10,000 = 25

Table 4.3

Coding Attribute Type Data

ATTRIBUTE DATA	Good	Bad
	+	-
	1	0

CLASSIFIED DATA	Excellent	Good	Poor	Bad
CODED DATA	1	2	3	4

Table 4.4

Attribute data may also be collected with a bar code label or a magnetic stripe that identifies the defect or characteristic. Coding schemes create enough flexibility to record defects by item, location, and cause. A defect-by-item scheme is shown in figure 4.3. One list of codes identifies the defects. Each separate code entry counts as a defect type. Rapid entry is achieved by passing a reader over the bar code when a defect is spotted. The sequence of codes is then sorted by a computer, and a control chart, histogram, or Pareto diagram can be formatted and printed. Defect-by-location and defect-by-cause check sheets can be simulated in essentially the same way. Separate code lists identify the defects, locations, machine workers, and times of day. Bar codes or magnetic stripes reduce dependence on an operator's command of the

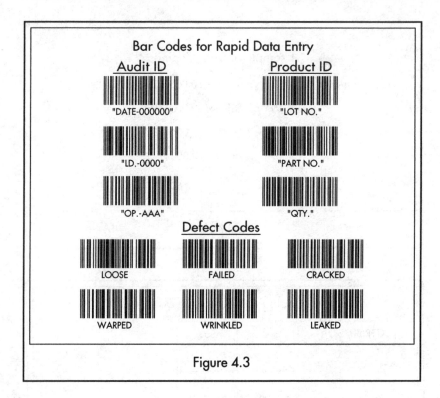

Figure 4.3

codes and the ability to key them correctly. The data is then transmitted to a computer for sorting by defects and other variables. The reports can also be set up to highlight unusual occurrences.

Negative numbers

Positive and negative numbers are used to distinguish between opposite qualities. Values above zero are positive and take the (+) sign. Values below zero are negative and use the (-) sign. The (+) and (-) signs are used as addition and subtraction and as signs of quality. When there is no sign, the (+) sign is understood.

Negative numbers are common in statistical data, particularly in the calculation of averages and ranges the construction of control charts. The steps for handling are as follows:

1. Negative numbers and then positive numbers are totaled separately. The results are then added. For example: average 2, 4, 6, -8, -7.

$$-8 + -7 = -15$$
$$2 + 4 + 6 = 12$$
$$\text{Sum of } -15 \text{ and } +12 = -3.$$

2. Then, the number of observations (n) in the sample is determined.

3. Last, the average is computed by dividing the sum of the observations by the number of observations.

$$\overline{X} = \frac{Sum}{n} = \frac{\Sigma X}{n} = \frac{-3}{5} = -.6$$

If the sum is a negative number, the average is negative also.

To add numbers of like signs, add the numbers and give the result the common sign. To add numbers of unlike signs (as in the above example), combine all positive and negative quantities, subtract the smaller from the larger total and give the results the sign of the larger combination.

Ranges are computed by subtracting the smallest observed value from the largest observed value. For example:

$$\text{Range} = (+6) - (-8) = +14$$
$$+6 + +8 = +14$$

Summarizing data

Data can be summarized using probability statistics. This reduces a large amount of data to a few statistics for analysis. *Statistics makes us appreciate with exactitude what intuitive minds feel by instinct.*

Populations and sampling statistics

Different symbols are used to distinguish between population and sampling data (table 4.5).

For populations, the Z-distribution is the standard, bell-shaped normal distribution (see Chap. 5).

The t-distribution is used when the sample size is less than 120. For samples that are larger than 120, the t-distribution becomes the same as the standard normal distribution of Z. The t-distribution is used in place of the Z-distribution because of the sampling variance's sensitivity to different values of n when n is less than 30 (see Chap. 5).

Bias

Some statistics, such as the average ($\overline{X}$) and the sample variance (s^2), are unbiased estimators. In this example, they are the mean (μ or $\overline{\overline{X}}$) and variance (μ^2). Others, like the range (R) and sample standard deviation (s), are biased estimators. Bias is a systematic error that contributes to the difference between a population mean (μ) of measurement or test result (X) and an accepted reference value (s):

	Mean	Size	Variance	Standard Deviation	Normal Distribution
Different Symbols for Population and Sampling Data					
Population	μ or $\bar{\bar{X}}$	N	σ^2	σ	Z
Sample	$\bar{X}$ or $\bar{Y}$	n	s^2	s	t

Table 4.5

$$\mu - ts$$
$$\mu - X$$

A good estimate of bias requires averaging a large number of test results representative of the population involved. In this instance, the standard deviation (σ) is, and a correction factor is necessary if the bias is to be removed. Frequently, bias is not a practical concern, particularly if the sample is large enough (over 30), if it is constant, and if all comparisons are made on the same basis. To reduce bias (when it is necessary) and to avoid an accusation of bias, random sampling should be employed. This is discussed in the next section.

Randomization

Randomization ensures that a sample is representative of the population and is free of biases due to undetected or unknown causes. It is necessary in data gathering and is an essential element in the design of experiments. The operational procedure assigns random numbers or uses a similar method for assuring that each unit has an equal chance of being selected.

Randomization may seem to be a bothersome and annoying procedure, but it can be carried out in a short time and is necessary to prevent biased data. It should never be excluded from statistical analysis. Successful data gathering depends heavily on spotting possible sources of bias and eliminating them with an ingenious, well-organized plan, before the sample is selected.

The following example of random sampling uses table A.8 in Appendix A, which features 800 random two-digit numbers from 01 to 99. Start with a desirable goal: in this case, selecting a sample of sixteen numbers from a lot of eighty. Any place on the table will be a good starting point to start choosing numbers from 01 to 80. A number that has already occurred must be discarded until 16 distinct numbers are selected. For example: 10, 16, 35, 15, 79, 61, 68, 70, 62, 25, 46, 58, 24, 39, 55, 72 represents a random sample of 16.

Many calculators and computers have random number generators that greatly simplify this process.

Replication

One of the chief sources of statistical power is the manner in which the samples are planned before the data is even collected. Since experimental error is almost invariably present, *replication* is used to increase the precision of effects estimates. In order to do this effectively, all elements contributing to experimental error should be included in the replication process.

For some experiments, replication may be limited to repetition under essentially the same conditions: using the same machine, a short time interval, or a common batch of materials. For experiments that call for general results, replication may need different but similar conditions: different machines, longer time intervals, or different batches of materials.

In some experiments, a *pseudo-replication* occurs when factors that have no effect (average or differential) are included in the experiment; two essentially identical brands, such as Coke and Pepsi, may be considered replicates rather than truly different versions.

In control charts, replication is achieved with rational subgroups of samples, which should be as free as possible from assignable causes. A series of samples will then show the differences in machine settings, batches of materials, and so forth.

In a production plant, gathering experimental data or making changes involves a large amount of supervisory manpower, raw materials, and overhead. So, to take more observations than necessary to be sure of significant results can be very expensive. With replication, accuracy becomes greater the more a process or experiment is repeated. The error of the average is inversely proportional to the square root of the number of observations:

$$\text{ERROR} = \frac{1}{\sqrt{n}}$$

Thus, averaging four samples brings the error of the average down to one-half of a single observation. The average of sixteen observations has one-quarter of the error of a single observation.

Frequency distributions

A *frequency distribution* represents a set of all values that individual observations may have and how often they occur in a database. When summarizing large masses of raw data, it is useful to distribute the data into categories, and determine the number of individuals belonging to each. The frequency distribution eventually becomes a graph, where related data has been grouped into intervals of equal size, called cells. Certain characteristics, such as the shape of the distribution, stand out more clearly in this format. Other characteristics, such as the most and least frequently occurring data values, can also be determined from a frequency distribution.

Directions for making a frequency distribution

To make a frequency distribution, follow these steps:

1. Select the correct number of cells. This is very important, for the data must resemble its distribution. If too few cells or too many cells are used, the picture that results is not very useful. Table 4.6 is a rule of thumb for determining the number of cells needed to construct a useful distribution.

Number of Cells

NUMBER OF VALUES	NUMBER OF CELLS
Under 50	5 to 7
50 to 100	6 to 10
100 to 250	7 to 12
Over 250	10 to 20

Table 4.6

2. After the number of cells has been chosen, use the following formula to determine the cell interval:

Frequency Distribution						
	Cell Midpoint	Cell Limits	Tally	Frequency		CUM %
LOW	8.3	8.1 - 8.5	III	3	3	2.97
	8.8	8.6 - 9.0	I	1	4	3.96
	9.3	9.1 - 9.5	IIII	4	8	7.92
	9.8	9.6 - 10.0	HH HH III	13	21	20.79
	10.3	10.1 - 10.5	HH HH HH HH III	23	44	43.56
MEDIAN	10.8	10.6 - 11.0	HH HH HH HH II	22	55/45	65.35
	11.3	11.1 - 11.5	HH IIII IIII IIII I	21	34	86.14
	11.8	11.6 - 12.0	IIII III	8	13	94.06
	12.3	12.1 - 12.5	IIII	4	5	98.02
HIGH	12.8	12.6 - 13.0	I	1	1	99.01
				Total = 100		
				Total + 1 = 101		

Figure 4.4

CELL INTERVAL =

HIGH VALUE - LOW VALUE
NUMBER OF CELLS

A cell interval is the distance between cell boundaries in terms of data units plotted. Cell boundaries are the end points of a cell; they constrain all values within the cell. It is customary for the cell boundaries to have one more significant figure (usually a 5) than the values being plotted. Table 4.6 is the database used to construct the frequency distribution in figure 4.4. For a database of 100 values, ten from table 4.7 were selected to determine the cell interval.

3. After determining the cell interval, the largest and smallest numbers in the raw data from the data base are used to find the range (difference) between the highest and the lowest values. For figure 4.4, the largest and smallest numbers from table 4.7 were 12.8 and 8.3.

$$\text{cell interval} = \frac{Range}{No.\ of\ cells} = \frac{12.8 - 8.3}{10} = .45$$

round off .45 to .5.

Sample Data

PART NAME (PRODUCT): Torison Bar
OPERATION (PROCESS): Induction Heat Treat - Gain in db
SPECIFICATION LIMITS: 12 db
OPERATOR: M. B. Cole
MACHINE: Meko #1
GAGE: Audiometer
UNIT OF MEASURE: dbA
ZERO EQUALS:

TIME	DATE	Meas. 1	Meas. 2	Meas. 3	Meas. 4	Meas. 5	SUM	AVERAGE $\bar{X}$	RANGE R	NOTES
1	3/31	11.1	9.4	11.2	10.4	10.1	52.2	10.4	1.8	
2	3/31	9.6	10.8	10.1	10.8	11.0	52.3	10.5	1.4	
3	3/31	9.7	10.0	10.0	9.8	10.4	49.9	10.0	.7	
4	4/1	10.1	8.4	10.2	9.4	11.0	49.1	9.8	2.6	1
5	4/1	12.4	10.0	10.7	10.1	11.3	54.5	10.9	2.4	
6	4/4	10.1	10.2	10.2	11.2	10.1	51.8	10.4	1.1	
7	4/4	11.0	11.5	11.8	11.0	11.3	56.6	11.3	.8	
8	4/4	11.2	10.0	10.9	11.2	11.0	54.3	10.8	1.2	
9	4/4	10.6	10.4	10.5	10.5	10.9	52.9	10.6	.5	
10	4/5	8.3	10.2	9.8	9.5	9.8	47.6	9.5	1.9	2
11	4/5	10.6	9.9	10.7	10.2	11.4	52.8	10.6	1.5	
12	4/6	10.8	10.2	10.5	8.4	9.9	49.8	10.0	2.4	
13	4/6	10.7	10.7	10.8	8.6	11.4	52.2	10.4	2.8	
14	4/6	11.3	11.4	10.4	10.6	11.1	54.8	11.0	1.0	
15	4/6	11.4	11.2	11.4	10.1	11.6	55.7	11.1	1.5	
16	4/7	10.1	10.1	9.7	9.8	10.5	50.2	10.0	.8	3
17	4/7	10.7	12.8	11.2	11.2	11.3	57.2	11.4	2.1	
18	4/8	11.9	11.9	11.6	12.4	11.4	59.2	11.8	1.0	
19	4/8	10.8	12.1	11.8	9.4	11.6	55.7	11.1	2.7	
20	4/8	12.4	11.1	10.8	11.0	11.9	57.2	11.4	1.6	

Table 4.7

4. A scale is constructed along the vertical axis to include the range of data and its boundaries. In one example (see fig. 4.5), a scale with ten (10) cells and a cell interval of .5 is constructed.

5. Next, determine the cell midpoint. The cell midpoint is the average of the two cell boundaries. It is customary to assign the value of the cell midpoint to all the observations in the cell.

6. Tally the individual values from the database (table 4.6) in the appropriate cells with a slash mark for each observation. The total used to determine the cumulative percent is calculated by adding the slash marks. Figure 4.4 shows the completed frequency distribution.

7. Add the number one (1) to the total (TOTAL + 1) to prevent the cumulative percentage from reaching 100 percent. This is done for two reasons. When data is plotted on probability paper, there is no place for 100 percent. Also, the data is taken from a sample and not the total population, which means the cumulative percent must be less than 100. One hundred percent of the data cannot be in a sample. Since the total in the example is 100, then 1 + 100 = 101. For figure 4.4, 100 divided by 101 = 99 percent.

Frequency histograms

A *frequency histogram* is a frequency distribution graph that features rectangles with bases equal to the cell interval and areas proportional to the frequencies. The bars represent the frequencies in each category and are used like the cells in a frequency distribution. When data is summarized in this form, it yields, at a glance, an effective picture of the data without sacrificing information. The following characteristics of a histogram should be noted:

- an estimate of the central tendency of the distribution is where most of the data is clustered. This is the average or mean and can be estimated roughly by eye;
- the median (50 percent point), or the value in the middle of the data, can be found by starting in one tail and counting to the middle observation.
- the mode (or modes) is the value that occurs most frequently;
- the range of the data is the difference between the largest and smallest values; and
- the shape of the distribution shows whether it is normal, skewed, or bimodal.

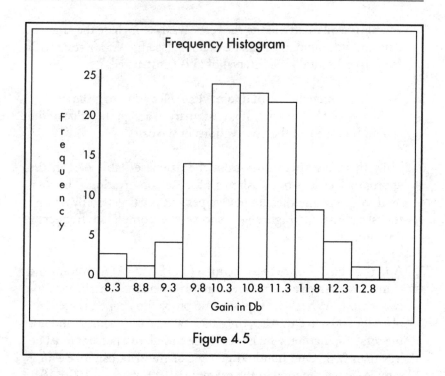

Figure 4.5

Figure 4.5 shows a frequency histogram. Note that the scale runs along the vertical axis. It is marked with the cell boundaries and labeled to include the range of the data.

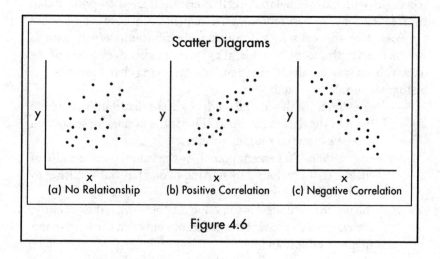

Figure 4.6

Scatter diagrams

The simplest way to study relationships (or *correlations*) is to plot a *scatter diagram* (*or scatter plot*). Scatter diagrams can determine if there is:

- a possible cause-and-effect relationship;
- a relationship between one cause and another; or
- a relationship between one cause and two or more other causes.

If no relationship exists, the plot will be amorphous. If a possible relationship exists, the dispersion will form a pattern with either a positive or negative slope.

Typical scatter diagrams are illustrated in figure 4.6.

Directions for making a scatter diagram

To construct a scatter diagram, follow these steps:

1. Obtain values for the variables (x) and (y) in pairs.

2. Collect twenty-five or more of these paired observations. For example, it is possible that there is a relationship between the hardness of a material and the strength of a part. To verify this, the hardness of twenty-eight pieces was measured, followed by the tensile strength of the same pieces. Table 4.8 has the set of paired observations used in figure 4.7.

3. Draw the horizontal axis (X) and the vertical axis (Y) for the scatter diagram. Determine a scale that covers the range of values for each of the paired observations (see table 4.7).

4. Plot the first paired observations as a single data point. Continue plotting until all data points are plotted.

5. Determine if a relationship or correlation exists by dividing the data into four equal sections. For example, using the data from table 4.8, divide the twenty-eight paired data points so that fourteen (one-half) are above the (Y) median and the other fourteen are below the median. Then divide the data on the (X) axis so that one-half of the points is to the left of the median and the other half is to the right of the median.

6. Count the number of paired observations in each of the four quadrants. These are observed data values that are to be used for a chi-square analysis.

Paired Observations for a Scatter Plot

Brinell Hardness	Tensile Strength (psi)	Brinell Hardness	Tensile Strength (psi)
240	122,100	236	117,300
251	120,200	260	126,000
247	123,100	238	118,200
255	125,800	250	122,600
240	118,300	259	120,200
247	119,200	251	117,000
252	125,600	256	125,800
238	112,900	244	123,300
254	121,800	247	114,600
239	118,400	256	126,100
245	114,700	250	123,800
255	124,300	243	120,400
250	119,100	260	127,200
242	117,400	253	122,700

Table 4.8

Completed Scatter Diagram

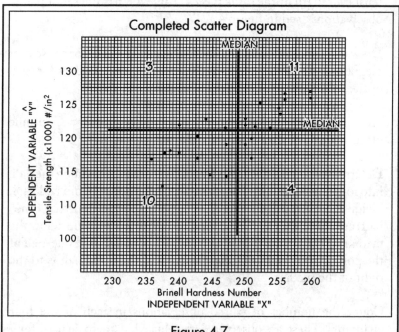

Figure 4.7

7. Analyze the data using the chi-square (x^2) distribution (see Chap. 6 for a full description of the analysis procedure). When there is no correlation, there will be an equal amount of paired data points in each section or quadrant. This is known as the null hypothesis (H_0). If a relationship does exist, the quadrants will have an unequal amount of data pairs resulting in a significant chi-square value.

Check sheets

Check sheets collect data in a form that may be used easily and analyzed quickly. A check sheet combines a tally sheet with a scatter diagram to create graphical diagram. It is particularly powerful when the check marks are made on a drawing or picture of the part. A scatter diagram (fig. 4.7) is fundamentally a check sheet of paired observations. A frequency distribution (fig. 4.4) is another example of a check sheet.

Check sheets determine graphically when and where problems are occurring in a process or product. To understand the location of a defect or problem, a transparent grid sheet is placed over a sketch or drawing of the product or process. The grid lines are drawn in evenly and at intervals close enough to isolate the defect.

Figure 4.8 shows a location grid on the door frame of an automobile used to find fitting defects. The locations of reported leaks around the door of a particular automobile model were marked with Xs. The accumulation of these marks on the drawing helped determine the problem areas. A tally was then made of the grid location of the reported leaks. See table 4.9.

The leaks were occurring at two specific locations (A and C) opposite each other. Upon investigation, it was discovered that the automobile assembly fixture was slightly out of alignment. The fixture

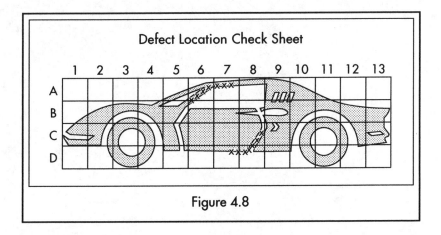

Figure 4.8

Defect Location		
Defect Location	Tally	Total
A 6	I I I I	4
A 7	I I I	3
C 8	I I	2
D 7	I I	2
D 8	I I I	3
	Total	14

Table 4.9

was adjusted, and the leaks were eliminated. This is an example of a matrix check sheet that isolated the problem and led to the locating of the problem in the manufacturing operation.

Conclusion

The most effective methods of data collection have been detailed in this chapter. They represent the second step in the plan for investigation laid out in Chapter 3. Chapter 5 discusses how to use the data collected by various means and how to interpret it. That is the third step in the plan.

Chapter 5

Data Analysis

"Man's mind cannot grasp the causes of events in their completeness, but the desire to find the causes is implanted in man's soul."

-Count Leo Nikolayevich Tolstoy

In this age of computers, which are fast and precise, the reams of data generated as computer printouts are overwhelming. Because there are so many of them, these printouts are filed and eventually thrown away when no one knows what to do with them.

Failure to use the data is a waste of basic information. If this data is used, it has often been found to be inaccurate because the method of collecting the data has been lost.

Examining samples from a population can produce information about the general nature of its characteristics while reducing the volume of data needed to be reviewed (fig. 5.1).

One reduction device is a frequency distribution histogram. When a sample is large (over thirty) the histogram will show the general nature of the corresponding distribution.

For a number of populations, many different curves or distributions arise. To analyze data on the basis of samples, an understanding of the types of distributions is necessary.

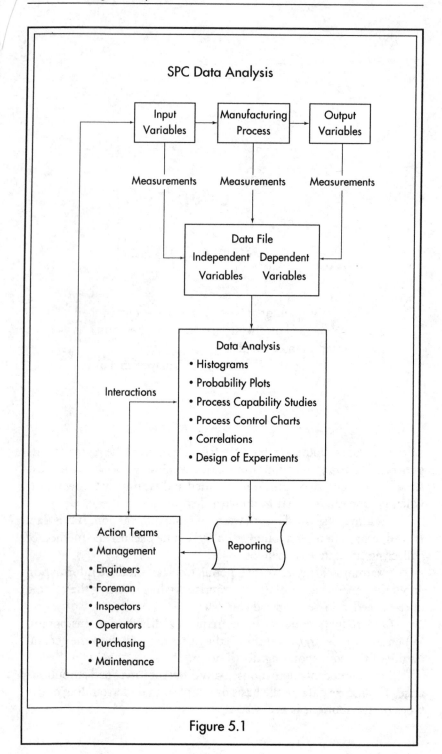

Figure 5.1

Five different distribution patterns are encountered in SPC. These distributions are:

- standard normal;
- binomial;
- Poisson;
- chi squared; and
- log normal.

The most important distribution in statistics is the normal distribution. Its graph, called the normal curve, is bell-shaped and describes the distribution of many sets of data that occur in nature and industry. A normal distribution has a specific shape and can be described by two statistics: the arithmetic mean of the distribution (μ) and the standard deviation (σ).

Figure 5.2 shows the normal bell-shaped curve in what statisticians agree is a standard form. The peak of the curve is directly over the

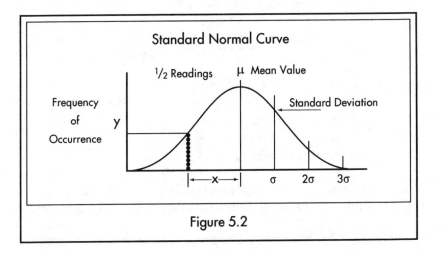

Figure 5.2

arithmetic mean (μ) of the measurements. The spread of the curve is specified by its standard deviation (σ). The greater the standard deviation, the broader the curve will be. A reduction in the base of the curve will cause an increase in height.

The peak of the curve is directly over the arithmetic mean of the measurements. It is useful to remember that the standard deviation is the distance from the mean to either of the two inflection points on the curve. The inflection point is the point at which the curve changes from convex upward to concave downward. The standard deviation is a special property of the normal distribution, because it is used to describe the shape of the distribution.

Distribution characteristics

Distributions have three characteristics that provide useful information:
- central tendency, or average;
- spread, or dispersion; and
- shape.

Central tendency

The *central tendency* is the area of distribution where data tends to cluster. The most common measures of central tendency are: mean, median, and mode.

The *mean* is commonly referred to as the average. A population mean is the total of the items or units under consideration divided by the number of them:

$$\frac{\Sigma X_i}{n} = \frac{Sum\ of\ Individual\ Values}{Number\ of\ Observations} = \overline{X}(\text{X bar})$$

$$= \frac{X_1 + X_2 + X_3 + X_4 + X_5}{n} = \overline{X},\ where:$$

Σ = a sum of a series of numbers,
$\overline{X}$ = sample mean (or estimate of the true population mean),
  i = number of observations or cases
      (or number of individual values to be added),
  n = total number of units in the sample.

When data is arranged in order of size, equal numbers of measurements are counted from either end of the series until a single value is left at the center.

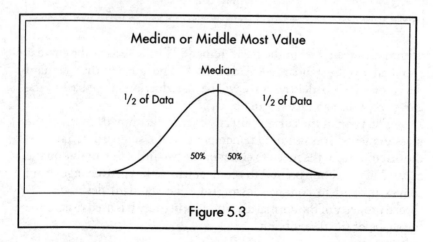

Figure 5.3

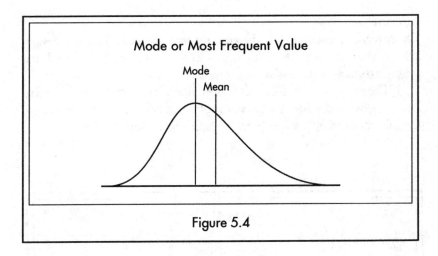

Figure 5.4

That is the median. If two values are left, the median is the average of the two. See figure 5.3.

The *mode* is the most frequent value of the variable. It can be identified as the maximum point on the distribution curve. In a skewed (not evenly distributed) distribution, the mode does not occur at the same point as the arithmetic mean. See figure 5.4. The mode can also help determine whether data is coming from two or more independent sources; if it is, the data may be bimodal or trimodal (fig. 5.5).

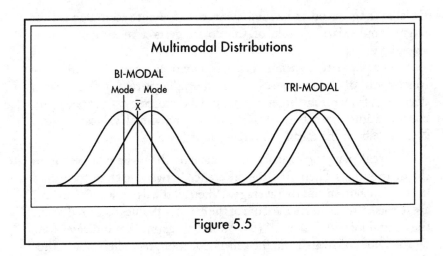

Figure 5.5

Spread

The spread, or dispersion, is the amount that the individual values vary around their average. Among the common measures of spread are *range, variance,* and *standard deviation.*

The *range* (R) (fig. 5.6) is the difference between the largest and the smallest observed value in a given sample. It is calculated by subtracting the minimum sample value from the maximum sample value:

$$Range = X_{max} - X_{min}.$$

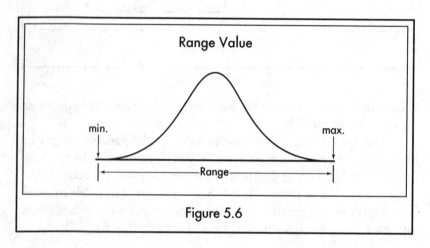

Figure 5.6

Because ranges tend to be inefficient, they are not recommended for large sample sizes. A rule of thumb suggests a maximum sample size of ten.

The population *variance* is a mathematical measure of variability (dispersion) of observations. It is based on the average of the squared deviations from the arithmetic mean and is denoted by σ^2. It takes into consideration the extent of a cluster of individual values (X_1, X_2, X_3,...X_N) about the mean or average value (μ or X).

To calculate the variance, first calculate the mean of all values in the distribution. Then, find the difference between each of the values in the distribution and this average. When all the deviations $\Sigma(X_i - \overline{X})$ are totaled, the result is a measure of the extent of values clustered about the central location value, $\overline{X}$ (fig. 5.7). This expression indicates how far the values of $\overline{X}$ are from the central location, regardless of whether the values are above or below the mean ($\overline{X}$).

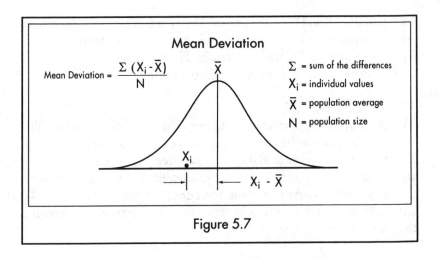

Figure 5.7

Mean Deviation= $\dfrac{\Sigma(X_i - \overline{X})}{N}$, where:

Σ = sum of the differences,

X_i = individual values,

$\overline{X}$ = population average, and

N = population size.

The sums of the mean differences are then squared. If they were not squared, this difference would always be zero. Squaring each deviation corrects this problem and still reflects the spread or dispersion of the values. The mean of the squared deviation from the arithmetic mean is the population variance: σ^2 (sigma squared):

$$\sigma^2 = \dfrac{\Sigma(X_i - \overline{X})^2}{N} \text{ , where}$$

Σ = sum of the differences;

X_i = individual observed values;

X = sample average; and

N = population size.

Standard deviation (σ) is the preferred dispersion measure. It is based on the spread of all the observations. Like range, it is expressed in the same measurements as the data values. However, 6σ is the range for 99.8 percent of normally distributed observations.

Shape

The third most important distribution characteristic is *shape*. The observed data may either be symmetrical or irregular about the mean. The normal curve is unimodal and symmetrical, so the mean, medium, and mode will be the same for the standard normal distribution. A mean change slides the curve right or left but does not alter its profile. A the standard deviation change widens or narrows the curve without changing the location of its center. ˙

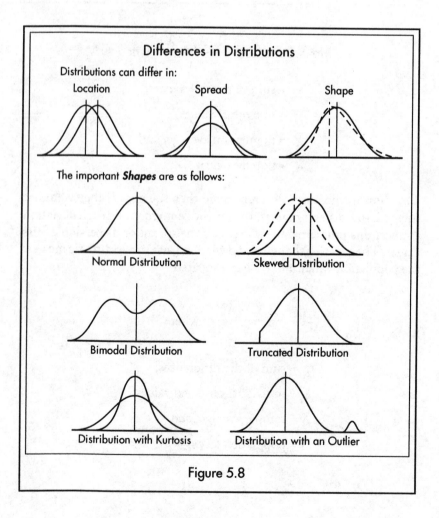

Figure 5.8

The important shapes are as follows (fig. 5.8):
- normal distribution;
- distributions that are symmetrical but not normal;
- distributions with various degrees of skewness; and
- distributions showing more than one mode.

Use of standard normal distributions

SPC usually involves measurements of independent variables: diameters, speeds, temperature, time, voltage, loudness, amounts, and so forth. All are quantities with an infinite number of graduations and will always vary according to chance and changing conditions, no matter how carefully measured. The measurements take on a continuous range of data values that are symmetrical and bell-shaped, extending infinitely in both positive and negative directions. These shapes tend to be normally distributed.

There are actually many different normal curves. Each is bell-shaped and can be mathematically defined. Each results from the point where the curve is centered and the areas to which it spreads out on either side. The equation is complicated. To use the standard normal distribution, the variable must be expressed in a standardized form. The mean (μ) should be equal to zero, and the standard deviation (σ) is equal to 1. Z is the distance from the population mean (μ) in the units of the standard deviation and is computed using the equation:

$$Z = \frac{x_i - \mu}{\sigma}, \text{ where:}$$

x represents any value in the population in various intervals of a normal distribution.

Tables have already been calculated from the equation. Table A.2 in Appendix A serves as a reference for the standard normal distribution (fig. 5.9) and is called the Z curve. The table gives a proportion of the total area under the portion of the curve from $-\infty$ to

$$\frac{X_i - u}{\sigma}.$$

X_i represents any desired value of the variable X. For example, suppose the mean of a distribution is 100 and its standard deviation = 10. What would be the probability that the variable (X) will be greater than 120?

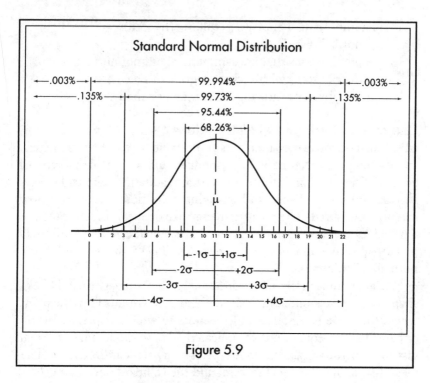

Figure 5.9

$$Z = \frac{X_i - u}{\sigma} = \frac{120 - 100}{10} = \frac{20}{10} = 2.00.$$

Locate Z = 2.00 in the standard normal distribution table (table A.2 in Appendix A). The probability of Z being greater than 2.00 is noted as .0228 or 2.28%.

The Z-distribution table gives the probability for only positive values of Z. Since the normal curve is symmetrical, negative values of Z are the same as the positive values (fig. 5.10).

Sampling distributions

Process control more often deals with process samples than with the universe population. However, the value of a sampling statistic is subject to variability and may result in different samples arriving at different conclusions.

Information about the long-run behavior of the process is needed to obtain a measure of reliability. This is obtained by applying probability concepts to a random sample. The random sample may be the units of material themselves or observations taken from them.

One measure of sample variability (dispersion) is the sample variance (s^2), which is the deviation from the arithmetic average squared

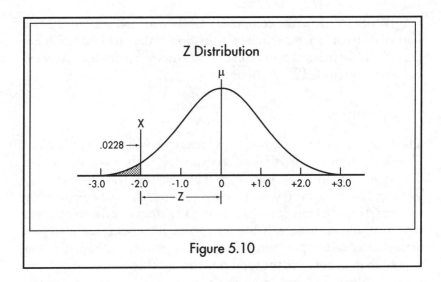

Figure 5.10

and divided by the degrees of freedom. s^2 should always be accompanied by its degrees of freedom. In this example, the degrees of freedom equal the sample size less 1. From a simple random sample, s^2 gives an unbiased estimate of the population variance (σ^2).

$$s^2 = \frac{(X_1 - \overline{X})^2}{n-1} \text{ or } \frac{(Y_1 - \overline{Y})^2}{n-1}, \text{ where:}$$

X_i = sample value,

$\overline{X}$ = sample average, and

n = sample size.

Unbiased means that the values of s^2 from all possible samples in the same population will average to (σ^2). The sample variance s^2 is the best estimate of σ^2 in the case of a normal population. The standard deviation is the square root of the sample variance (s^2).

Student's t-distribution

While the Z-distribution is used for an analysis of a population distribution, the t-distribution is used for sample data. Just as there are many different normal distributions, there are many different t-distributions. S (standard deviations of the sample) is used instead of (σ^2). This substitution employs the sample standard deviation as an estimate of the population's standard deviation, and the sample size must be taken into account. Each corresponds to a different sample size.

When X^1, X^2,....X^n comprise a random sample size, n, from a normal distribution, the sample distribution of the standard variable is the ratio of the mean of a particular sample subtracted from an expected value to its corrected sample size:

$$t = \frac{X_i - \overline{X}}{s / \sqrt{n}}$$

The t-distribution is expressed in Appendix A (table A.3) for a given number of degrees of freedom equal to n - 1. It is known as student's t. This name has an interesting story. The equation was discovered by William Gosset, a thirty-two-year-old research chemist employed by Guiness Ale. The famous Irish brewery had a strict rule that employees were never allowed to publish their discoveries. Because of the importance of this statistical computation, however, Gosset was granted permission to publish it, but only on the condition that he not reveal his name. He chose "student" and reported it in Volume 6 of *Biometrika Journal* in 1908. His real name was released in later years, but it is his pseudonym that remains one of the most renowned names in statistics.

When student's t-distribution is compared with Z-distribution, the result may resemble the graphs in figures 5.11 and 5.12.

Important properties of the t distribution are as follows:
- the t-distribution is more spread out than the Z-distribution;
- as the sample size (degrees of freedom) increases, the spread of the corresponding t curve decreases; and
- as the sample size (degrees of freedom) becomes increasingly large, the corresponding t-curve approaches the Z-curve. For samples above 120, they are identical.

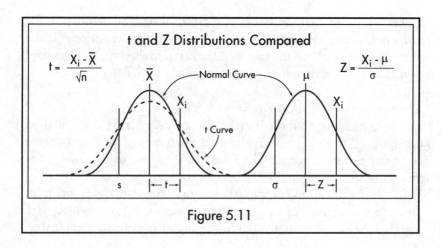

Figure 5.11

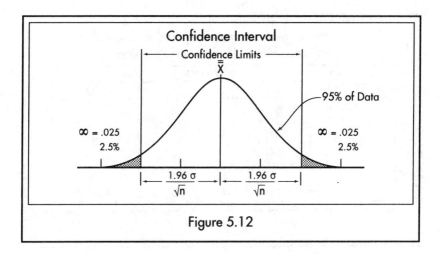

Figure 5.12

Sample size

"What I tell you three times is true."

-Anonymous

The methods of inductive statistics enable us to learn only about the overall properties represented by the sample and not about characteristics of specific values.

The *sample* size (n) is the number of data values, collected in such a manner that all combinations of (n) under consideration have an equal chance of being selected. It should be large enough to allow an estimate made from them to approximate a normal distribution. Too large a sample is a waste of resources, and too small a sample reduces the usefulness of the results.

As can be seen from table 5.1 and figure 5.13, there should be a minimum sample of eight. This is about the smallest number from which an inference about a distribution may be made. To compare one sample with another, sixteen samples are needed. Eight observations on the old process and eight on the new can detect a shift in one standard deviation (fig. 5.13).

A sample of thirty will suffice for a good average and variance.

The sample size necessary to estimate the standard deviation can be determined with certain precision. If it can be expressed as a percentage (P%) of the true (unknown) standard deviation, then the data in table 5.1, which was determined from the t-distribution, may be used.

| \multicolumn{4}{c}{Number of Measurements Required to Establish the Variability with a Stated Precision} |
| ALLOWED % DEVIATION | CONFIDENCE COEFFICIENTS | | |
	.90	.95	.99
.50	6	8	14
.40	8	12	21
.30	16	20	37
.20	32	45	83
.10	135	190	330
.05	540	770	1325

Table 5.1

Increasing the sample size increases the precision of the sample data. However, an increase in precision is not directly proportional to the sample size, but only to the square root of the sample.

Increasing the sample sizes, therefore, is an inefficient and uneconomical direction to take simply to gain more information. Instead, process changes should be introduced, and one should take samples of them. (See Chap. 7.)

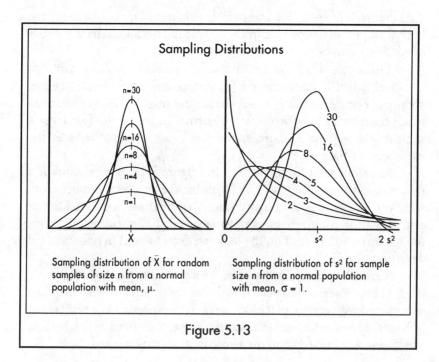

Sampling Distributions

Sampling distribution of $\bar{X}$ for random samples of size n from a normal population with mean, μ.

Sampling distribution of s^2 for sample size n from a normal population with mean, $\sigma = 1$.

Figure 5.13

Confidence interval estimates

A sample average will seldom be exactly the same as the lot, or population, average. It should be fairly close, with an interval that brackets the population mean. For example, if samples did contain the true mean 95 percent of the time, then there is a 95 percent confidence level. The common confidence levels commonly used are 99 and 95 percent, which correspond to $\alpha = .01$ and $\alpha = .05$. It is assumed that the data of interest has a normal distribution in the population.

Confidence limits are the end points of the sample statistics that, with a specified *confidence coefficient* $(1-\alpha)$, include the population average and standard deviation.

Interval estimates may be two- or one-sided. A two-sided interval estimate brackets the mean from above or below, while a one-sided estimate is limited on either the upper or lower side.

Histograms

Basic to data collection and the analysis of data is the histogram. A histogram made up from a sample will show some of the characteristics of the parent distribution. This is useful for:

- suggesting the shape of the parent or population distribution;
- indicating certain discrepancies or peculiarities in the data;
- comparing a collection of data with other samples of data;
- comparing data with specifications; and
- forming a basis for probability plots.

Histograms should not, however, be used to draw general conclusions. Data can change with time. Histograms should be used with control charts to determine if the data is changing or is in control.

There are several types of specialized histograms:

- stem-and-leaf displays;
- two-sample frequency histograms;
- hanging histobars;
- box-and-whisker plots; and
- multiple-box and-whisker plots.

Stem-and-leaf displays

Stem-and-leaf displays preserve all of the original sample data. This is a simple method of organizing numerical data and can be done with pencil and paper. A stem-and-leaf display shows the range of similar data values, where these values are assembled, and how balanced they are, while highlighting gaps in the data and identifying outliers. The advantage to the stem-and-leaf frequency histogram is that the sample data is retained for each distribution.

Each number in the data set is divided into two parts. The leading digits form the stem, and the trailing digit or digits form the leaf. Once a choice of stem values is made, the leaf values are listed out to one side.

Figure 5.14 is an example of a stem-and-leaf display. The data is taken from table 4.6, which represents 100 data values. For example, with this DB data, the stem consists of whole numbers and the split was at the decimal point, with the leaf being the decimal values.

Stem and Leaf Histogram

```
 3      80 | 344
 1      8* | 6                    8* | 6 represents 85.6
 4      90 | 4445
 9      9* | 67888899
23     100 | 00001111111112222224444
20     10* | 55556667777788888899
26     110 | 00000011122222223333444444
 9     11* | 56668899
 4     120 | 1444
 1     12* | 8
```

Figure 5.14

Computer programs that arrange and plot data on command are available. For more extensive applications of stem-and-leaf histograms, see John W. Tukey, *Exploratory Data Analysis*, 1977.

When plotted back-to-back, stem-and-leaf displays can be used for a visual comparison of two data sets. In figure 5.15, the two data sets contain many of the same values, but it is obvious that the information differs. The stem-and-leaf-distributions may be tested for significant differences by counting the number of data points that exceed the paired distributions. In figure 5.16, the exceeding values are identified with a plus symbol. The statistical test for exceeding values, and the probability of their occurrence, may be obtained from table 5.2.

Two-sample frequency histograms

Two-sample frequency histograms are useful comparison methods. Two histograms are plotted back-to-back to illustrate their differences. Figure 5.15 illustrates.

Data in a two-sample frequency histogram may also be interpreted with a stem-and-leaf display (see fig. 5.16).

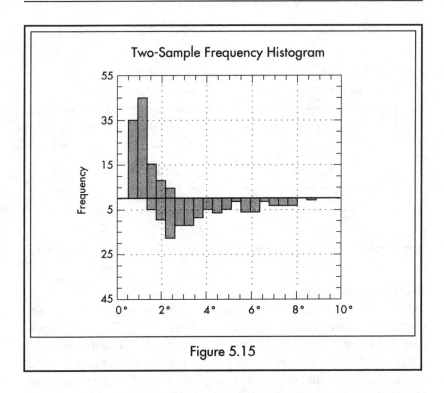

Figure 5.15

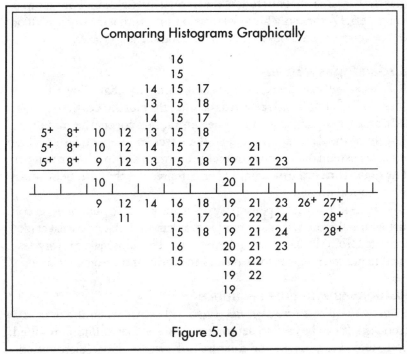

Figure 5.16

Exceeding Values Probability		
No. of Exceeding Values	Probability of Occurrence	
7, 8, 9	1/20	5%
10, 11, 12	1/100	1%
13 & over	1/1000	0.1%

Table 5.2

Hanging histobars

In a histobar, the bars are suspended from the best-fitting normal distribution curve and not from the horizontal axis. The vertical axis then represents the degree of deviation from the expected frequency. The bottoms of the bars should be randomly scattered in a narrow band around the horizontal line at zero. Any pattern that emerges indicates that the base line does not follow the normal distribution. For example, in figure 5.17, there is a bimodal distribution, with a major distribution at 2 to 3 degrees and another minor distribution at 6 to 9 degrees.

Box-and-whisker plots

A *box-and-whisker plot* is a graphical summary that allows for the detection of outliers and skewed data, since the plot divides the data into four areas of equal frequency. The central box covers the middle 50 percent of the data values between the lower and upper quarters. The whiskers extend out to the extremes (minimum and maximum values). The central line is the median, and an X represents the arithmetic mean or numerical average.

Box-and-whisker plots provide a means of comparing several histograms. Figure 5.18 is an example of a multiple box-and-whisker plot compiled from the data used in figure 5.16. This plot makes it very easy to compare the range of the data and the mean and median locations.

Histograms with time sequence

A histogram with a time sequence (fig. 5.19) shows when the data was produced. It can be used to determine if a skewed distribution resulted from tool wear, freaks, or mixtures of two or more distributions

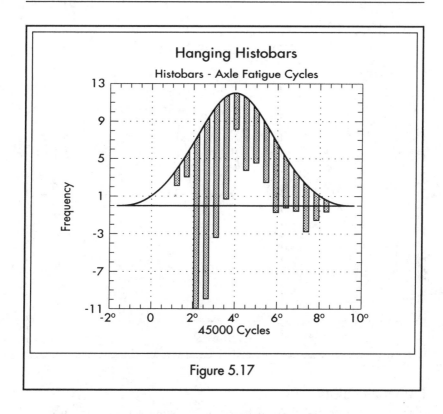

Figure 5.17

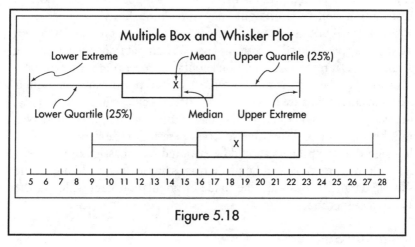

Figure 5.18

occurring in time. Control charts, however, would be better for this application because they also evaluate the range of the data as well as the means.

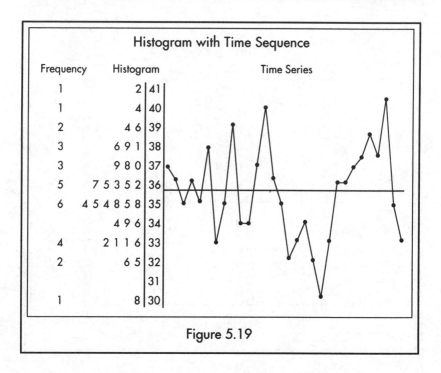

Figure 5.19

Probability plots

A probability plot (fig. 5.20) is based on the supposition that a normal distribution will be a straight line when probability paper is used. Process distributions can be tested for normality on probability paper, which is a special kind of graph paper. When data does not plot as a straight line, it reveals information about the process data and can show reasons why the data is not normal. A curved plot means the data is skewed; curves that are offset or have different slopes indicate bimodal or even multi-modal data. When data is multi-modal, there may be a number of significant variation sources operating at the same time. This is a sign that the process should be explored for an assignable cause.

Probability plots can be made from sample data, utilizing all the sample data. They can also be made of population data utilizing the histogram method.

Important Statistics

Important statistics may be obtained from a probability plot. These include:

- the process average;
- the estimated standard deviation;

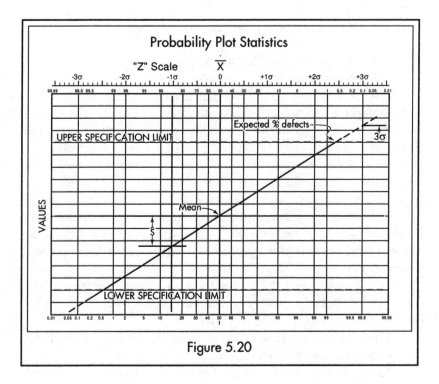

Figure 5.20

• outliers; and
• comparisons.

The *process average* (x) is the point where the "line of best fit" crosses the 50 percent line, or the mean of the distribution. The *estimated standard deviation* (s) is the difference between the point where the line of best fit crosses the 16 percent (or the 84 percent) and the 50 percent line. The slope of the line is a measure of variance; the steeper the line, the greater the variance. Points that vary greatly from the line of best fit are *outliers*.

Probability plots may be used to compare distribution processes, compare the probability distribution with specification limits on probability paper, and analyze data from a sample or batch process.

Probability plots using sample data

A probability plot is one of the most important techniques for doing a sample analysis or for obtaining data from a batch process. This is because it automatically compares the sample data with a normal distribution. Some processes that can be analyzed by a sample probability plot are:

• a batch process;

- a product coming from only one supplier;
- a product that has lost its source;
- initial production that has no prior history;
- process capability studies; and
- a process too rapid for in-process data collection.

Comparing probability distribution with specification limits determines not only the capability of the process, but also how well-centered the distribution is. If the process is not centered, the probability plot may be used to decide the amount of adjustment required to center the distribution. Often, it is apparent that a small shift in the process will improve the process capability dramatically.

If changes were made after the initial data was plotted, these may be plotted on the same sheet for comparative purposes. This is an efficient method for tracking process capability for continuing process improvement.

To construct a probability plot, there are a number of steps. First, collect the data to be analyzed. The size of the sample should be between ten and sixty individual values.

Then, rank the data from the smallest to the largest, the most positive to most negative, and the earliest to the latest.

Assign the rank 1 to the smallest observation, rank 2 to the second smallest, and so on, assigning "n" to the largest observation in the sample. In table 5.3, "n" = 30.

Calculate a probability plotting position for each observation, from its rank (i) and the total number of samples (n) or use table A.1 in Appendix A:

$$P_1 = \frac{100(i - .05)}{n}, \text{ where:}$$

P_1 = the plotting position, and

i = 1, 2, 3...n.

For a chosen distribution, label the data scale to span the data. Then, plot each sample observation against its plotting position on probability paper. In figures 5.21 and 5.23, the bobbin data is plotted on normal probability paper. Either the top (fig. 5.21) or bottom (fig. 5.23) probability scale may be used.

Using a transparent ruler or French curve, draw in a "line of best fit." This is when half of the points on one side fit a best line to the data

| Typical Data for a Probability Plot |||||||||
| Data on Temperature Rise of Bobbins |||||||||
Obs.	Rank i	100(i-0.5)/n	Obs.	Rank i	100(i-0.5)/n	Obs.	Rank i	100(i-0.5)/n
44.7	1	1.7	45.9	11	35.0	46.5	21	68.3
44.7	2	5.0	45.9	12	38.3	46.5	22	71.7
45.0	3	8.3	45.9	13	41.7	46.5	23	75.0
45.1	4	11.7	46.0	14	45.0	46.7	24	78.3
45.3	5	15.0	46.0	15	48.3	46.7	25	81.7
45.3	6	18.3	46.1	16	51.7	47.0	26	85.0
45.4	7	21.7	46.2	17	55.0	47.2	27	88.3
45.7	8	25.0	46.2	18	58.3	47.4	28	91.7
45.8	9	28.3	46.3	19	61.7	48.1	29	95.0
45.8	10	31.7	46.4	20	65.0	50.6	30	98.3

Table 5.3

(see figs. 5.21, 5.23, 5.24, and 5.25). The emphasis should be given to the points between 10 and 90 percent, with the data points evenly distributed above the ruler or curve. Add the upper and lower specification limits, then a median line, and identify the standard deviation point.

The probability plot may now be examined and analyzed for the desired information.

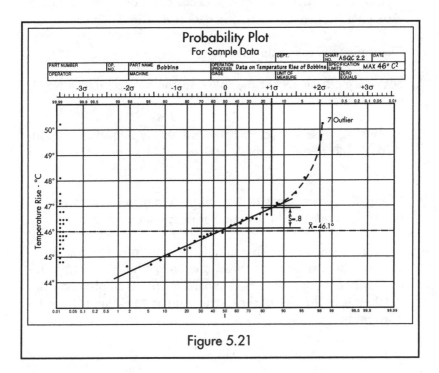

Figure 5.21

Probability plots using histogram data

When a large amount of data (50 items or more) is being collected for analysis, a histogram or frequency distribution is an easier method of data collection.

To use histogram data to construct a probability plot, these procedures should be followed:

1. Collect the data to be analyzed. (The data for this example is from table 4.6 and fig. 5.22.)

2. Identify the minimum and maximum values.

3. Group the data into cells. To determine the number of cells to be used, refer to table 4.5. In this example, a database of 100 was grouped into 10 cells.

4. Determine the cell size by subtracting the minimum value from the maximum value and divide by the number of cells determined previously.

5. To determine the cell end points, round off the cell size into a logical value. Make sure these values do not overlap. Record the cell sizes on a frequency distribution data sheet (fig. 5.22).

6. Determine the cell midpoints.

7. Make a mark, or record the value, in the tally column for each sample value.

8. Record the total for each tally column in the frequency column. Add all the frequency totals and record the grand total. Add 1 to it.

9. Obtain the percentages by subtracting 0.5 from the frequencies and divide this number by total number in the sample plus 1.

10. Obtain the values for the cumulative percents. Do this from the top down to a mode value, and then from the bottom up to a mode value. This is a crucial step; it emphasizes any differences in the tails of the distribution and highlights bimodal distributions.

Frequency Distribution Data Sheet

PART NUMBER 125B	PART NAME Amplifler	OP NO. 120	PART NAME Southern Electric		
CHARACTERISTIC Gain In Db			ENGINEERING SPECIFICATION 11 ± 2 Db		
DATE	REMARKS				

CELL SIZE FROM - TO	MID-POINT	TALLY	FREQ.	PERCENT	CUM.
12.6 - 13.0	12.8	I	1	.49	.49
12.1 - 12.5	12.3	IIII	4	3.47	3.96
11.6 - 12.0	11.8	ᵗᴴᴸ III	8	7.43	11.39
11.1 - 11.5	11.3	ᵗᴴᴸ ᵗᴴᴸ ᵗᴴᴸ ᵗᴴᴸ I	21	20.30	31.69
10.6 - 11.0	10.8	ᵗᴴᴸ ᵗᴴᴸ ᵗᴴᴸ ᵗᴴᴸ II	22	21.29	52.98 / 63.38
10.1 - 10.5	10.3	ᵗᴴᴸ ᵗᴴᴸ ᵗᴴᴸ ᵗᴴᴸ III	23	22.28	41.09
9.6 - 10.0	9.8	ᵗᴴᴸ ᵗᴴᴸ III	13	12.38	18.81
9.1 - 9.5	9.3	IIII	4	3.47	6.43
8.6 - 9.0	8.8	I	1	.49	2.97
8.1 - 8.5	8.3	III	3	2.47	2.47

TOTAL 100
TOTAL + 1 101

$$\% = \frac{FREQ. - .50}{TOTAL + 1}$$

Figure 5.22

11. Construct a value scale of the vertical axis (Y) on the probability sheet, using the minimum amount of scale space possible. The value scale should include both the upper and lower specification limits, as well as the full range of the individual values. Where possible, the range values should be roughly centered. (See figs. 5.21, 5.23, and 5.25.)

12. Using the percent scale at the top and bottom of the probability sheet, plot the cum percentages.

13. Using a transparent scale or a French curve, draw in a line of best fit. A normal distribution will plot as a straight line. If the data does not plot as a straight line, it indicates a disturbance to the distribution.

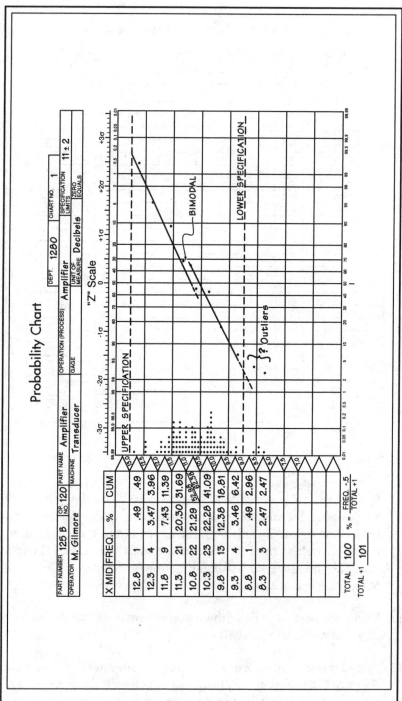

Figure 5.23

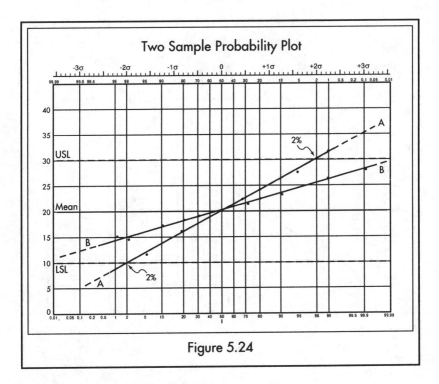

Figure 5.24

Two-sample probability plots

Since the mean and standard deviation on a probability plot are visibly apparent, two or more processes, along with specifications, may be compared easily on one probability sheet. The slope of the probability plot is evidence of the variance. In figure 5.24, both distributions have the same mean, but A's slope is steeper than B's. Thus, process B is better.

Interpreting probability plots

Distributions are used to interpret probability plots (see fig. 2.9) some comparisons. The standard normal distribution, which is a straight line, is the benchmark.

A *truncated distribution* results when the data has been cut short in some way. Products that have been inspected and removed for being out of tolerance after inspection would show a truncated distribution. A *skewed distribution* is curved, not straight. A positive skewed distribution (fig. 5.25) is convex with a long curve to the right. A negative skewed distribution is concave with a curve to the left. Tensile test breaking strengths have a characteristic of this type. A *bimodal distribution*, two overlapping distributions, is characterized by two relatively straight sections that are offset and connected by a jog, which

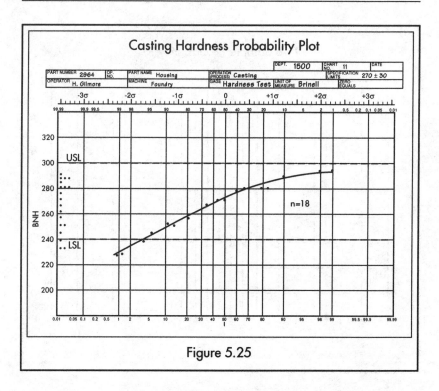

Figure 5.25

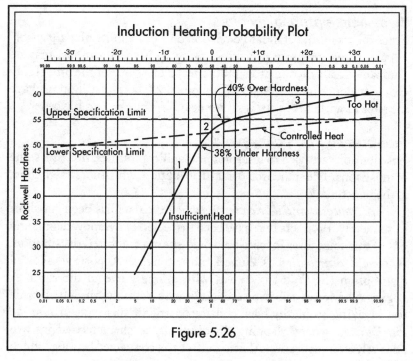

Figure 5.26

is a brief, abrupt change in direction. The degree of offset is evidence of the difference in the means of the distributions. This may occur within a single sample (see fig. 5.23). The sources may be two different machines, operators, shifts, or materials. Determining its cause and bringing it under control would result in an improvement of the overall variation in the process. Two jogs in a probability plot indicate a trimodal distribution with data from three different spindles or operators.

It is not unusual to find a mixture of curves in a distribution. Figure 5.26 illustrates a problem that occurred in an induction heating machine that resulted in three curves. These curves, however, indicated the problem and directed a way to the solutions. The part of the curve at area 1 indicating a low Rockwell hardness (Rc less than 50) suggested that a portion of the product output received insufficient heat. Area 3 shows a high Rockwell hardness (Rc greater than 55), indicating too much heat. Area 2 indicated a transition from being too soft and not having enough heat to being too hard with too much heat. Subsequent investigation found the temperature sensor to be faulty. The sensor was replaced with a more accurate infrared heat sensing device.

Specifications
A probability plot not only allows analysts to compare process distributions at once on the same probability sheet, but it also may be used to compare these distributions with their specifications. See figs. 5.24, 5.25, and 5.26. The latter two illustrations show effective ways of tracking process improvements, for they indicate how both the process mean and its variance relate to the specifications.

Defects
Probability plots may also be used to detect expected defects in percentages (see figs. 5.25 and 5.26) at both the high and low ends, without any calculations. These percentages are determined by the point where the line of best fit intersects with the specification limits and are read directly from the top or bottom percentile scale.

In the example in figure 5.25, it is readily apparent that the distribution is not centered. This results in a higher percentage of values (4%) being out of specifications on the lower side.

Outliers
Probability plots can help analysts draw conclusions about an extreme value or an outlier. An *outlier* is a data point that is far removed from the majority of measurements or reflects a gross error in data collection. This problem often occurs when data is supposedly taken under the

same conditions, but there are one or two observations that are widely different (an *outlier* from the rest). The problem that confronts an analyst is whether the outlier should be kept (as a bad unit) or discarded (as bad data).

If an observation is "discarded," it is rejected in computation only. Every observation should always be documented. A careful investigator will want a record of the "rejected observations" in his or her careful search for an assignable cause.

Figure 5.21 shows that more than half the data is above the specification 46 degrees Celsius maximum. It has also identified an outlier at 50.6 degrees Celsius. In this example, the outlier appears to be part of a skewed distribution toward the higher temperature. Temperature, then, probably is an assignable cause.

Control charts: time series distributions

A control chart is essentially a picture of a sampling distribution. It is a most effective method of handling SPC problems, controlling quality during manufacturing, and indicating when action should be taken to prevent quality problems from occurring. It evaluates the range of data as well as the means. Time series control charts allow a sufficient opportunity for unforeseen or unintentional changes (or variables) to appear in the process.

A control chart is a graphical record of data samples taken in a time sequence. Such data fluctuates naturally in a pattern that has a bearing on any problem, series of events, or manufacturing situation. When plotted on a chart, the numbers always form a zigzag. Testing the pattern can lead to information about the process. An important feature of a time series is that it allows unanticipated and unintentional changes (or variables) to appear in the process. Any and all factors that could occur at any time may be taken into consideration and identified at the time they occur.

To exercise control over a process, limits assist in judging the significance of the fluctuating variations. These control limits are placed so that a plotted point falling outside them during a process indicates a variation that should be investigated. By making use of certain equations, derived from statistical laws, it is possible to calculate limits for any given pattern. If a pattern is natural, its fluctuations will fit within these limits. If a pattern is unnatural, its fluctuations will not fit these limits (AT&T, 1956).

Principal kinds of control charts

A *range* chart documents the difference between the highest and lowest data values in a small sample. *Averages* chart the average values of samples. These are the two most sensitive types of data.

Percentages (p-charts) chart the proportion of bad (or other attributes) to the total number of units (areas of opportunity) sampled. A p-chart is the second most sensitive chart for identifying processes. It is limited by the fact that it has only one pattern and thus cannot identify causes accurately. However, its pattern is similar to those of X and R charts (cycles, trends, freaks, gradual changes, sudden shifts, and so forth). Such patterns can be related to one's knowledge of the process variables.

Moving range ($M\overline{R}$) is a chart of individual measurements with control limits based on the moving range.

Process capability charts are used to discover whether a process is behaving naturally or unnaturally to collect process-operation information, as well as investigate and eliminate unnatural behavior. When these unnatural disturbances are eliminated, the remaining, or natural behavior, is called process capability.

Process capability charts should be placed along production paths. These can be determined from the flow process charts. Splitting data between machines, shifts, operators, operations, and other controlling elements creates ideal comparisons.

A *process control chart* is based on the process capability study. It is used to maintain a predetermined distribution over a long period of time, and therefore serves as a continuing capability study.

The process control chart is maintained by and benefits the operating organization. It is used to prevent defects, detect shop troubles at the source, and maintain a smooth, running process, all of which result in cost reduction.

Since process control charts are a time series analysis of a process, it is important that they be placed as close as possible to the beginning of the process that is to be controlled. These initial processes include raw materials, purchased parts, castings, forging, vendor service, initial machining, or initial process steps. Control charts that come at the end of a process are not much use as "control charts"; instead, they are simply final inspection charts. The emphasis of process control is "doing it right the first time" and this should be translated as "as early in the process as possible" (AT&T, 1956).

The two most sensitive kinds of data, ranges and averages, are mapped together to form an $\overline{X}$ and R chart. In this manner, they become a powerful diagnostic tool (AT&T, 1956). Instructions for constructing and making calculations for $\overline{X}$ and R charts and the symbols used may be found in Appendix E.

Control chart for attributes
Attributes exist in SPC and other administrative processes, so the analysis techniques are useful in many applications. Data gathered for manufacturing or management summary reports are often in attribute form and can benefit from a control chart analysis. As mentioned earlier in this chapter, a p-chart is often used for plotting attributes.

A p-chart does not require any actual measurements such as diameter, length, or size. One need only count the number of the two possible outcomes and change this count to percentages. Measurements of this type, which depend on counting, are called attribute measurements. The attribute-type charts are also called percentage charts. The p means proportion, and it represents the proportion of the attribute count as compared to the total.

A p-chart is not as sensitive as $\overline{X}$ and R charts, which depend on the analyst's knowledge of the variables in the process. On the other hand, a p-chart often has the advantage of using records readily available in the shop. p-charts may also be used for:
- measuring characteristics for which it is difficult or impractical to obtain variables measurements;
- studying an entire assembly process with an overall chart; and
- measuring the effectiveness of changes, corrections, or improvements made as a result of other studies (AT&T, 1956).

Instructions for constructing and making calculations for p-charts, as well as the symbols used, may be found in Appendix F.

Charts for individual measurement
Charts for individual measurements have their control limits based on a moving range. The principal kinds of data for which this chart should be used are:
- accounting figures of all kinds. This includes shipments, efficiencies, absences, losses, inspection ratios, maintenance costs, accident reports, and records; and
- production data such as temperatures, pressures, voltages, humidity, conductivity, furnace heat, gas compositions, and the results of chemical analysis.

In all of these cases, only one number is available to represent a given condition.

In a moving range chart, $M\overline{R}$ represents moving range, or the difference between successive pairs of numbers in a series. Each individual number is used in calculating two of the moving ranges. MR is the average of a series of moving ranges (AT&T, 1956). Instructions for constructing a chart with moving range limits are in Appendix G.

Analyzing control charts

When a series of sample values is gathered together instead of being mapped in sequence, they form an up-and-down distribution pattern.

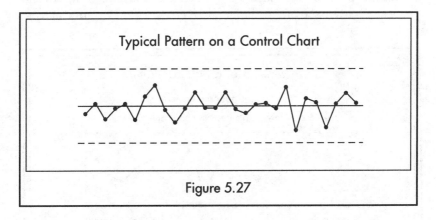

Figure 5.27

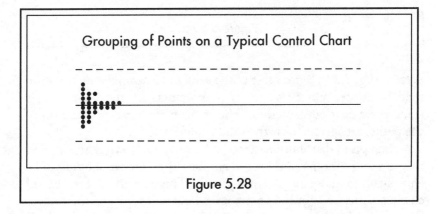

Figure 5.28

An example of this is shown in figure 5.27. When the plotted points were grouped together at one end of the chart, they formed a distribution like that shown in figure 5.28. In other words, when all control chart data is collected, the result is a normal distribution pattern.

The irregular, up-and-down pattern formed by points on a control chart can be classified as "natural" or "unnatural." Making this classification involves a look at each point to see whether it is part of an unnatural pattern, and marking the point with an "X" if it reacts to the visual inspection. Analysts should practice the tests in Appendix H until they can apply them automatically while glancing at a control chart pattern. All control charts would he marked immediately in accordance with these tests. Failure to mark the Xs on a control chart may result in its incorrect interpretation (AT&T, 1956).

Characteristics of a natural pattern

A natural pattern fluctuates randomly, and the points follow no particular system or order. Three characteristics make up a natural

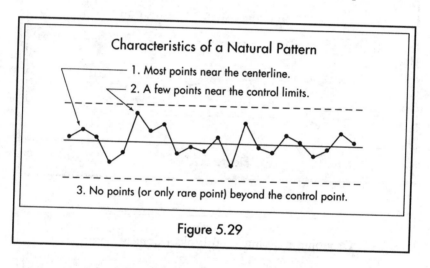

Figure 5.29

pattern (fig. 5.29). Since most of the values in a sampling distribution tend to cluster about its center, it is natural for the points to be somewhere near the centerline. This is the first characteristic: most of the points are located near the solid centerline.

Sampling distributions tend to be reasonably symmetrical, and therefore it is also natural for the number of points on one side to be equivalent to those on the other side. This is part of the second characteristic: a few of the points spread out and approach the control limits on each side of the centerline. These "tails" may extend as far as +/- sigma. An occasional point may reach the control limits. None of the points, except in a few rare instances, exceeds the control limits.

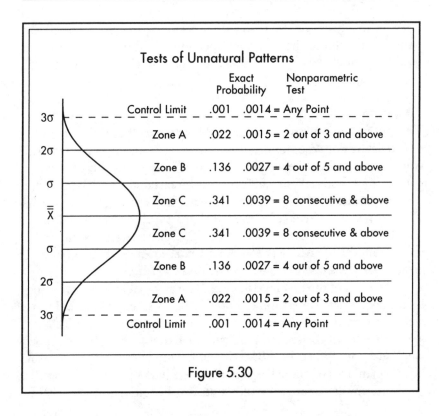

Tests of Unnatural Patterns

	Exact Probability	Nonparametric Test
Control Limit	.001	.0014 = Any Point
Zone A	.022	.0015 = 2 out of 3 and above
Zone B	.136	.0027 = 4 out of 5 and above
Zone C	.341	.0039 = 8 consecutive & above
Zone C	.341	.0039 = 8 consecutive & above
Zone B	.136	.0027 = 4 out of 5 and above
Zone A	.022	.0015 = 2 out of 3 and above
Control Limit	.001	.0014 = Any Point

Figure 5.30

Figure 5.29 illustrates these characteristics. If any are missing, the pattern becomes unnatural. Probability calculations indicate the proportion of natural points that fall near the centerline, near the control limits, and so forth. Figure 5.30 shows these probabilities.

Characteristics of an unnatural pattern

Many unnatural patterns can be recognized on the control chart. Unnatural patterns either fluctuate too widely or not enough. They also may fail to balance themselves around the centerline. The absence of one or more of the characteristics of a natural pattern always results in an unnatural pattern.

The presence of points outside the control limits is *instability*. An absence of points near the control limits is *stratification*. The up-and-down variations are small when compared to the width of the sample control limits. The sampling is done systematically: two or more different distributions are represented. An absence of points near the centerline is a *mixture*. Too many points fall near the control limits (AT&T, 1956).

Formal tests are also used for the scientific interpretation of the patterns. These may be found in Appendix H. When these tests are used, they should be applied to every control chart to ensure that all will be interpreted similarly.

Tests for instability

The most important tests for unnatural patterns are those for instability, which determine whether the cause system is changing. The standard control chart test (Shewhart, 1931) uses 3 sigma as a criterion for control limits and lack thereof. The tests for instability are based on the 3 sigma limits. Each sigma forms a zone that is one sigma in width. These zones are referred to as the one sigma zone, the two sigma zone, and the three sigma zone. See figure 5.30. This criterion should strike an economical balance between the net consequences of two types of error: *error of the first kind* and *error of the second kind*.

An error of the first kind is looking for assignable causes when no such causes exist. For example, a point falls outside the control limits when, in fact, there has been no change in the process (fig. 5.31).

An error of the second kind is not looking for assignable causes when such causes do exist. For example, a point falls within the control limits when, in fact, there has been a change in the process (fig. 5.32).

Information about instability tests may be found in Appendix H.

Inconclusive patterns

Occasionally, a pattern may not appear unnatural at the time it is plotted, but the trends of the points indicate that it might become so. This type of pattern is often identified as inconclusive. Figure 5.33 illustrates.

Analysts working with an inconclusive pattern have two choices: they may ignore the pattern until it is marked. The other, preferable option is to obtain more data immediately and allow the pattern a chance to complete itself to see if the cause system has really shifted.

In a process capability study or a designed experiment, often it is important to discover as soon as possible whether the pattern is going to react. In such cases, analysts try at once to obtain sufficient data so that they may complete or refute the test. For example, if the next point plotted in figure 5.33 completes the pattern, then the cause system has changed. If the next point breaks the pattern, then there is no evidence of a change. In these cases, the pattern itself has indicated how much data is needed to reach a conclusion (AT&T, 1956).

Error of the First Kind

Figure 5.31

Error of the Second Kind

Figure 5.32

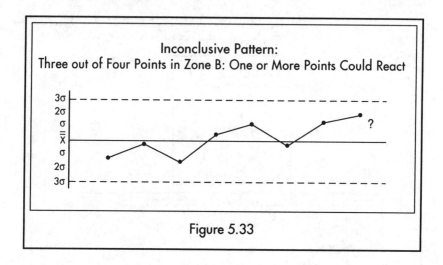

Figure 5.33

Other unnatural patterns

The processes of recognizing other unnatural patterns differ from instability tests. Both halves of the control chart are considered together in looking for the patterns shown below. These patterns are marked with circled Xs to distinguish them from the patterns of instability.

Stratification exists when fifteen or more consecutive points fall in Zone C, either above or below the centerline (fig. 5.34). *Mixture* exists when the chart shows eight consecutive points on both sides of the centerline and none falls in Zone C (fig. 5.35). A *systematic variable* in the process is indicated by a long series of points that are high, low, high, low without any interruption in this regular sequence (fig. 5.36)

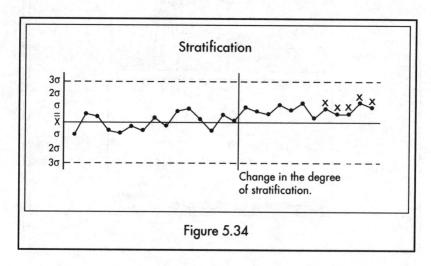

Figure 5.34

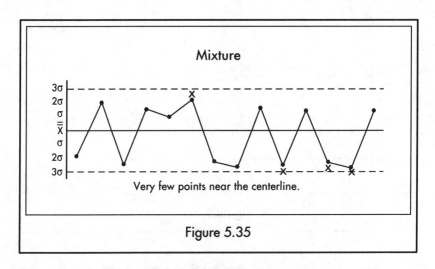

Figure 5.35

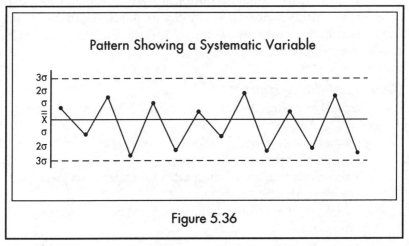

Figure 5.36

Trends may be indicated by either marks on one side of the chart followed by marks on the other or a series of consecutive points without a change in direction (fig. 5.37).

Meaning of the R-chart

The R-chart shows uniformity or consistency of the process. If the R-chart is narrow, the product is uniform. If the R-chart is wide, the product is not. If the R-chart is erratic, something is operating on the process in a nonuniform manner.

Machines in good repair and processes in good condition tend to produce uniform products. Carefully trained operators also aid in product uniformity.

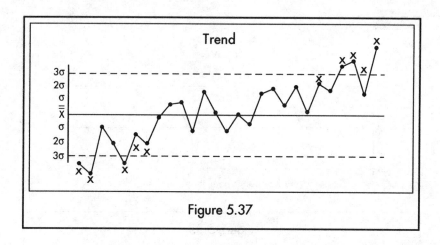

Figure 5.37

When an R-chart is out of control, analysts should look for poor repair or poor maintenance if it is a machine controlled-process. They should look for new operators or something disturbing the operators if it is on operator-controlled process.

Meaning of the $\overline{X}$-chart

The $\overline{X}$-chart shows where the process is centered. Processes are ordinarily centered by:

- a machine setting;
- some other process adjustment;
- the characteristics of the particular material or piece part being used; or
- a bias or change in technique on the part of an operator or inspector.

If the $\overline{X}$-chart is natural, the center of the process is not shifting. If the $\overline{X}$-chart shows a trend, the center of the process is moving up or down gradually. If the $\overline{X}$-chart is erratic and out of control, something is changing the center rapidly and inconsistently.

$\overline{X}$-charts can also be affected by out-of-control conditions on the R-chart. If the $\overline{X}$-chart and R-chart are both out of control, analysts should first examine the R chart for the causes affecting it (AT&T, 1956).

To find the relationship between the process and the specification limits, make sure that both the $\overline{X}$ and the R chart are in control and follow the steps in Appendix I.

Meaning of a p-chart (or other attributes chart)

A p-chart shows proportion: the proportion of product is classified as defective if it is a "percent defective" chart, the proportion classified as good if it is a "yield" control chart, and so forth. When a pattern changes a p-chart, it means there is a change in proportion.

On a percent defective control chart, a change in level may mean either the percentage of bad product is increasing or decreasing, or there is a change in what is being identified as defective. Both of these possibilities should be considered when a p-chart is interpreted.

If the p-chart is defective, look for causes that come and go spasmodically. Poorly trained operators and poorly controlled piece parts are two of the most common causes. An erratic p-chart is frequently a sign that further process controls are needed (AT&T, 1956).

Meaning of a chart of individual measurements

A chart of individuals should always be examined first for trends, which will resemble those on an X-chart and mean the same thing. Are the fluctuations narrower or wider? The fluctuations show uniformity or consistency in much the same manner as those on an R-chart.

Does the pattern stray far from one of the control limits? This may indicate that the distribution is "blocked" on that side.

Finally, analysts should look for any obvious peculiarities in the pattern such as cycles or "bunching." Knowledge of the process will identify the probable causes for such peculiarities (AT&T, 1956).

Notes on control charts

A control chart is a working document. When an out-of-control point is identified, it should be noted on the chart. Reason(s) for the out-of-control condition(s), the change(s) made, and action(s) taken should be identified with the out-of-control point or points. To do so is part of analyzing the chart and determining the cause of the out-of-control condition(s).

Attribute process control

Numerical measurements of the characteristics being controlled are necessary for $\overline{X}$ and R control charts. There are processes for which it is desirable to make only simple measurements in the form of attributes, rather than variables. An example is a product classified only as over or under a specified limit such as go/no-go gauging. In the case of attribute process control, it is the number of items that are over or under the limit that is important.

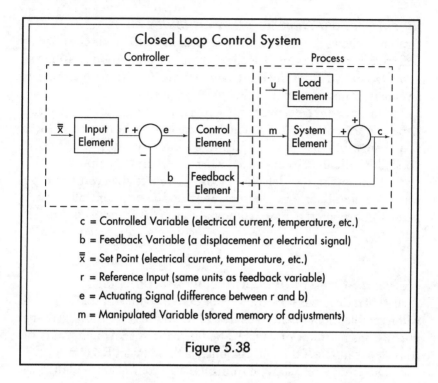

Figure 5.38

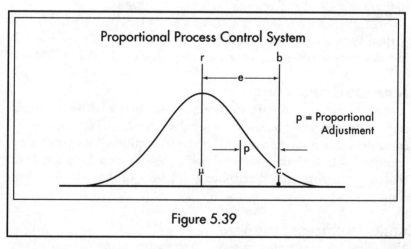

Figure 5.39

Normally, a feedback process control system includes a means of measuring the controlled variable's value. It compares it with the desired value, determines the deviation, and produces a proportional counteraction that will maintain control. A proportional controller will not work with attribute processes; instead, an incremental controller is used. An incremental adjustment is made, in place of a proportional adjustment, to a process deviation. See figures 5.38 and 5.39.

An attribute control system is based on the probability that an attribute will fall in one of three classes where the number of attributes in each class is:

$$n_x + n_y + n_z = n_{XYZ}.$$

The narrow limits dividing the three classes are represented by $t\sigma_1$ and $t\sigma_2$.

The incremental adjustment eliminates the need for a proportional sensitivity calculation. An incremental control system need only detect whether the product is outside the tighter limits (a gauge with tighter tolerance than a normal or process go/no-go gauge). If the product or process is above the narrow limits, then a

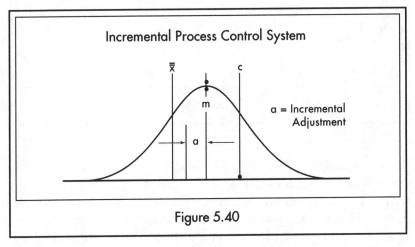

Figure 5.40

fixed incremental amount is subtracted from the process mean. In other words, if the attribute is greater than $t\sigma_1$, then the adjustment will be $n_{xa(-)}$. If it is below the narrow limits, then a fixed incremental amount is added to the process mean: $n_{ya(+)}$. See figures 5.40 and 5.41. The process mean is the variable being manipulated by an incremental controller in incremental amounts rather than proportional amounts.

If the attribute falls within $t\sigma_1$ and $t\sigma_2$, there will be no process adjustment. It is the number of parts above and below the limits that provide the feedback control. When the process is in control, these incremental adjustments to the process will balance thanks to the normal distribution's symmetry. If a real change takes place in the

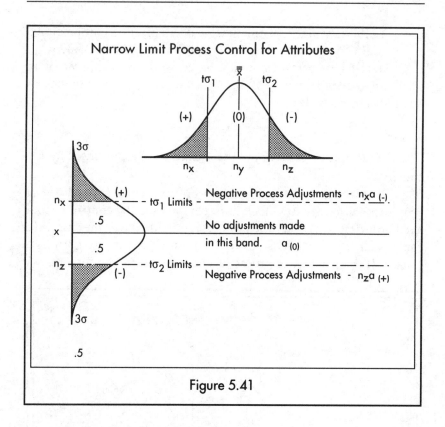

Figure 5.41

process or product, then there will be more adjustments in one direction than the other. The sum total of the correcting adjustments will bring the process back in control. Not making adjustments within the narrow limits brings fewer disturbances to the process than a proportional controller, which would be making adjustments on all detected values.

The statistical justification for an incremental control system is that when a process shifts, an imbalance is created. This is detected more quickly with narrow limits than with limits set at 3σ or greater. Adjustments are made before the process starts to produce units outside the specification. Further, there is no valid reason for making control adjustments on values within the narrow gauge limits; the variation in those limits is random for a stochastic process. Not making adjustments in this band (which would be the majority of the time) creates further disturbances in the process. An incremental feedback controller with narrow limits, then, is as good, if not better, than a proportional control device for manufacturing processes and would work just as well on variables data (Brewer, 1963).

Conclusion

In this chapter, the various ways of using charts and graphs to analyze data have been demonstrated. The five distribution patterns encountered in SPC and their uses were identified. Most importantly, the ways that graphical representations can be used to interpret and analyze data were discussed. In the next chapter, the use of statistics to interpret data will be introduced. This is also part of the third step of the plan for investigation.

Chapter 6

Statistical Analysis

This chapter concentrates on the application of the statistical techniques introduced in Chapter 5 and how they are used. Process improvement is covered in detail in Chapter 7.

Various distributions are used as references for making decisions and determining the significance of an event subject to chance variations. It is important to understand what the various distributions are and how they are used.

Continuous distribution: standard normal distribution

A continuous distribution is one of two main types. Its variable may take on all values in an interval and is represented by a smooth curve (see fig. 5.9). In the other main type, which will be discussed later in this chapter, the variable takes on only discrete values.

A continuous distribution has a specific shape and is described by the equation for the standard normal distribution:

$$Y = \frac{e^{-\frac{(x-\mu)^2}{2\sigma^2}}}{\sigma\sqrt{2\Pi}} \text{, where:}$$

e = 2.718,

σ = standard deviation (sigma) of the distribution,

μ = mean of the distribution,

X = measurement on the horizontal axis (Z score values), and

Y = height of the curve for an X value.

In many types of statistics and probability, the area under the curve described by the above equation must be known. These values are give in table A.2 in Appendix A

Use of continuous normal distribution

The mean and the standard deviation of the continuous normal distribution are particularly useful in both applied and theoretical statistics. They are easy to use because their properties have been thoroughly investigated. The continuous normal distribution was discussed in Chapter 5 in terms of the standard variable $Z = (x_i - \mu)/\sigma$. This transformation, or change, of variable permits the use of one table for all normal distributions despite their different means and variances. However, the variable must be expressed in a standardized form so that the mean is equal to zero and the standard deviation is equal to one. For this standardization, use

$$Z = \frac{x_1 - \mu}{\sigma}$$

The normal curve ($\mu = 0$ and $\sigma = 1$) is called the standard normal, or Z, curve (fig. 6.1), where Z is the continuous random variable for the standard normal distribution. Table A.2 in Appendix A gives the area under the Z curve both to the left of and to the right of different values along the measurement axis. See Chapter 5 for an example of the use of table A.2 to determine the probability of Z.

One of the most useful applications of the continuous normal distribution is the sampling distribution of the average. The sampling distribution of random sample averages will be normally distributed.

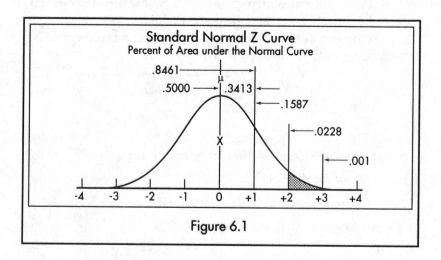

Figure 6.1

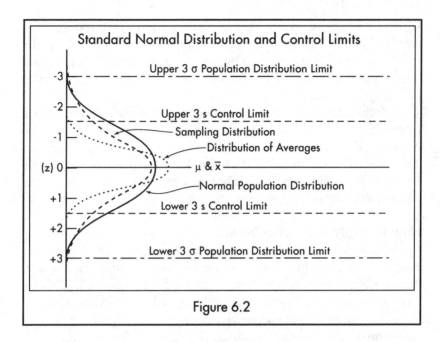

Figure 6.2

This is the foundation on which the control chart rests and is illustrated in figure 6.2.

The sampling distribution (student's t-distribution)

The sampling distribution is mathematically related to the population distribution from which the samples came. It determines the:
- the standard deviation of the sample;
- the mean or average; and
- approximates the normal curve, even when the parent distributions are irregular or skewed.

This relationship can be expressed as follows:

$$\sigma_{\bar{x}} = \frac{\sigma_\mu}{n} = \frac{R}{d_2} \text{ , where}$$

d_2 is a correction factor in the chart for ranges.

In statistics, this is called the central limit theorem. In essence, it asserts that when a sample size is large, probabilities involving the average (X) may be computed exactly as described by the standard normal distribution.

The t-distribution is similar to the Z-distribution. Both are symmetric about the zero mean. When a sample size is greater than 120, the two distributions become the same. In fact, the bottom line of the t-distribution table (see table A.3 in Appendix A) for degrees of freedom above 120 is equal to the Z values. Not surprisingly, then, the interpretation of student's t-distribution is the same as the Z-distribution. The larger the t value, the greater the assurance that the difference between x and X is significant.

Both distributions are bell-shaped; however, the t-distribution is more variable, because its values depend on the change in the statistics X and s. Z values depend on the more stable mean and standard deviation. Also the variance of the t-distribution depends on the sample size (see fig. 5.10). The t-distribution may be used as a reference distribution when s^2 is unknown.

How to apply the t-distribution

A production machine has been turning out twenty-five pieces per minute. In an attempt to increase its output, an adjustment is made to the machine. In three short test runs, the machine turns out 27, 29, and 28 parts per minute. Does this represent a significant increase or is it more likely the result of chance variation?

The production run average is: $\overline{X} = 25$.

The sample run average is:

$$\overline{X} = \frac{27 + 29 + 28}{3} = \frac{84}{3} = 28.$$

The sample standard deviation of the three sample measurements is:

$$s^2 = \frac{\Sigma(x - \overline{X})^2}{n-1} = \frac{(27-28)^2 + (29-28)^2 + (28-28)^2}{3-1} =$$

$$= \frac{(-1)^2 + (+1)^2 + (0)^2}{3-1}$$

$$= \frac{1+1}{2}$$

$$= 1$$

$$s = \sqrt{s^2} = \sqrt{1^2} = 1$$

$$t = \frac{X - \overline{X}}{s/\sqrt{n}} = \frac{25 - 28}{1/\sqrt{3}} = \frac{3}{.577} = 5.19$$

Degrees of freedom= n - 1 = 3 - 1 = 2.

According to table A.3 in Appendix A, for 2 degrees of freedom and a two-tailed level of significance, the t value, 5.19, lies between .05 and .02 or about .02. This means that the difference would occur two times in 100, a number too great to be assigned to chance. It can be concluded that the change was real and not simply a chance variation.

Interpretation of significance level

As indicated in the previous example, significance levels are usually interpreted as follows:

- .05 is a warning that something has likely occurred;
- .01 indicates that the event is statistically significant; and
- .001 indicates that the event is highly statistically significant.

Discrete distributions

Discrete distributions are characterized by values that are clearly separated from one another. Their variables can be divided into individually discrete classes. An example is a frequency distribution where the number of cases is counted.

Of the five distribution patterns encountered in SPC (see Chap. 5), three are discrete:

- binomial;
- Poisson; and
- chi-squared.

Binomial distributions

A binomial distribution provides the statistical basis for the p-chart (see Chap. 5). It deals with the number of times a defect (or characteristic) occurs. It can have only two possible outcomes (good/bad, pass/fail, yes/no), with one outcome having a probability designated as p, and the other, a probability of q. Thus, the probability of not having a defect or characteristic is $p + q = 1$ or $q = 1 - p$.

A binomial distribution (fig. 6.3) has an approximate normal distribution. The greater the value of n, the better the estimate of the binomial made from a normal curve. However, a normal curve is always a symmetrical distribution, whereas a binomial is symmetrical only in the special case where $p = 1/2$.

A binomial distribution may be used to fit a set of data classified in two groups. When only one sample size, *n*, is available, the binomial distribution indicates how well the unknown parameter (p) is estimated by the sample proportion (x/n). A binomial distribution may also be the basic distribution in attribute sampling and serve as the basis for a p-chart. It can be used to construct control charts for attribute-type data.

	Binomial Distribution	
Sample Size	Probability	Significance of Occurrence
n = 1	(.5)	.5
n = 2	(.5) (.5)	.25
n = 3	(.5) (.5) (.5)	.125
n = 4	(.5) (.5) (.5)(.5)	.063
n = 5	(.5) (.5) (.5) (.5)(.5)	.031
n = 6	(.5) (.5) (.5) (.5) (.5) (.5)	.015
n = 7	(.5) (.5) (.5) (.5) (.5) (.5) (.5)	.008
n = 8	(.5) (.5) (.5) (.5) (.5) (.5) (.5) (.5)	.004
n = 9	(.5) (.5) (.5) (.5) (.5) (.5) (.5) (.5) (.5)	.002
n = 10	(.5) (.5) (.5) (.5) (.5) (.5) (.5) (.5) (.5) (.5)	.001

Figure 6.3

These statistics are as follows:

$$\text{mean } (\mu) = \overline{p}$$

$$\text{standard deviation (s)} = \sqrt{\frac{\overline{p}(1-\overline{p})}{n}}$$

$$\text{control chart limits } (\pm 3\sigma) = \pm 3\sqrt{\frac{\overline{p}(1-\overline{p})}{n}}, \text{ where}$$

n is the number of samples.

Poisson distribution

When the proportion of defects (or characteristics) is small (less than 5) and the chance that they will occur is large, the binomial distribution approaches an important limiting distribution known as the *Poisson distribution*. A Poisson distribution is defined by the following properties:
- the number of characteristics occurring in one time interval (an hour, day, or week) or a specified region (a line segment, piece of material, area, or volume) are independent of those occurring in any other separate interval or region;

• the probability of a single characteristic occurring during a very short time interval or region size should not depend on the number of characteristics occurring outside the interval or region; and
• the probability of more than one characteristic occurring in such a small time interval or falling in such a small space or region is negligible.

The number of X characteristics in a Poisson distribution is called a *Poisson random variable*. The probability of this variable's representing the number of characteristics occurring in a given time interval or specified region is:

$$\text{probability} = \frac{\mu^x e^{-\mu}}{X!}, \text{ where}$$

$$e = 2.71828,$$

μ = average number of characteristics occurring

in a given time or specified region, and

X = Poisson random variable (e.g., 0, 1, 2, . . .).

Probability of a Traffic Ticket		
Number of Tickets per Person	Probability of X Tickets $p = \dfrac{\mu^X e^{-\mu}}{X!}$	
0	$e^{-\mu}$	= .3679
1	$\dfrac{\mu e^{-\mu}}{1!}$	= .3679
2	$\dfrac{\mu^2 e^{-\mu}}{2!}$	= .1840
3	$\dfrac{\mu^3 e^{-\mu}}{3!}$	= .0613
4	$\dfrac{\mu^4 e^{-\mu}}{4!}$	= .0153
5	$\dfrac{\mu^5 e^{-\mu}}{5!}$	= .0031
	Total	= .9995

Table 6.1

Traffic ticket example

For example, a Poisson distribution could be used to analyze the accumulation of traffic tickets. An average driver has had at least one traffic ticket. Assuming that he or she is law abiding, this situation has the properties of a Poisson distribution because:

- the number of traffic tickets acquired in one month is independent of tickets in any other month or location;
- the probability of a single ticket being issued in a month is proportional to the length of time and does not depend on the number of tickets occurring outside this time interval; and
- the probability of more than one ticket being issued in a month should be negligible (table 6.1). Therefore, the average number of tickets is 1.

Thus, the probability of one ticket is:

$$\mu = 1$$

$$e^{-\mu} = e^{-1} \quad = 2.718$$

$$e^{-\mu} = 1/e^{-\mu} \quad = .3679$$

$$X! = 1! \quad = 1$$

$$p = \frac{\mu^x e^{-\mu}}{X!} = \frac{(1)(.3679)}{1} = .3679$$

The accumulation of four tickets results in the loss of a license. This loss implies that a driver with four tickets differs significantly from a driver with one ticket. Based on the Poisson distribution, if the average distribution of the tickets per person is equal to 1, then an individual has the probability of .3679 or a little over one-third of a chance of receiving a ticket. The probability of a person having four or more tickets is .0153, which is less than 2 percent. Thus, the driver with four tickets is in the minority and must be doing something wrong.

Industrial example

An example of a Poisson distribution being used in manufacturing involves comparing the actual number of parts with X defects to the number of defects expected. The latter is determined by Poisson distribution calculations.

This comparison is made in table 6.2, while figure 6.4 is a graphical Poisson plot of the number of defects. The actual number of

Probability of Defect

Number of Defects "x"	Actual Number of Parts with x Defects	Probability of Defects $p = \dfrac{\mu^x e^{-\mu}}{x!}$		Expected Number of parts with x Defects $p = 50$
0	21	$e^{-\mu}$	$= .4070$	20.35 or 20
1	18	$\dfrac{\mu e^{-\mu}}{1!}$	$= .3663$	18.31 or 18
2	7	$\dfrac{\mu^2 e^{-\mu}}{2!}$	$= .165$	8.2 or 8
3	3	$\dfrac{\mu^3 e^{-\mu}}{3!}$	$= .049$	2.4 or 2
4	1	$\dfrac{\mu^4 e^{-\mu}}{4!}$	$= .011$	.6 or 1
	Total 50			Total 50

Table 6.2

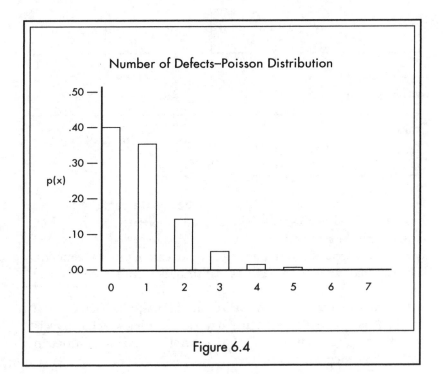

Number of Defects–Poisson Distribution

Figure 6.4

defects found were 45 out of a lot of 50 parts. This is an average of .90 or about 1 defect per part, when:

$$\mu = .90,$$

$$e^{\mu} = e^{.90} = 2.4569, \text{ and}$$

$$e^{-\mu} = e^{-.90} = 1/2.4569 = .4070.$$

Poisson distribution relationship to C and U charts

C and U control charts measure attributes and are based on the Poisson distribution. It is assumed that the possibility for defects and characteristics occurring on a part or a process is great, but the chances of a particular defect or characteristic appearing at any one area are small.

The following shows the relationship of Poisson statistics to C-chart and U-chart statistics:

	Poisson	C-Chart	U-Chart
Mean	$\bar{\mu}$	$\bar{c}$	$u = \dfrac{\Sigma \bar{c}}{\Sigma n}$ $(c = no.\,of\,defects)$ $(n = sample\,size)$
Variance (σ^2)	$\bar{\mu}$	$\bar{c}$	$\dfrac{\bar{u}}{n}$
Standard Deviation	$\sqrt{\bar{\mu}}$	$\sqrt{\bar{c}}$	$\sqrt{\dfrac{\mu}{n}}$
Control Chart Limits	$\pm 3\sqrt{\mu}$	$\pm 3\sqrt{c}$	$\pm 3\sqrt{\dfrac{\bar{u}}{n}}$

Both C and U charts are appropriate for the same basic data situations. The U-chart should be used if the sample includes more than one unit or if the sample size varies from period to period. This type of chart often is applied to computers and other complicated assemblies. The mean and variance of the Poisson distribution both have the value u or c.

Using Poisson distribution data from table 6.2, which features the number of days (frequency) in a fifty-day period during which X defects occurred on a manufactured part, the C-chart (fig. 6.5) might be set up with the following equations:

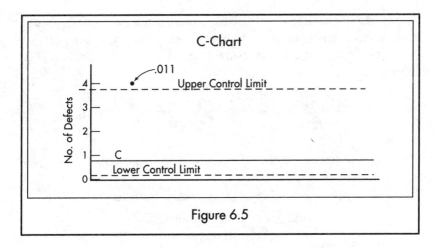

Figure 6.5

$$\mu = \frac{\Sigma fx}{n} = \frac{(21)(0) + (18)(1) + (17)(2) + (3)(3) + (1)(4)}{50}$$

$$= \bar{c} = \frac{45}{50} = .90$$

$$\sigma^2 = \bar{c} = .90$$

$$\sigma = \bar{c} = \sqrt{.90} = .949$$

$$\text{Control limits} = \bar{c} \pm 3\sqrt{\bar{c}}$$

$$\text{Upper control limit} = \bar{c} + 3\sqrt{\bar{c}} = .90 + 3\sqrt{.90} = 3.746$$

$$\text{Lower control limit} \quad \bar{c} - 3\sqrt{\bar{c}} = .90 - 3\sqrt{.90} = 0$$

Hence, four defects in a 50-day period would be highly significant (p = .011). The Poisson distribution nearly fits the actual data. However, no control limits have been given. Since the control limits are +3 σ from the center line, both sides of the center line can be divided into three sigma zones. Hence, the rules for out of control points for C and U charts are the same as those for X and R charts.

Chi-square distribution

The *chi-square (χ^2) distribution* is usually used to determine whether the actual frequency of an event that has occurred differs significantly from what was expected. The number of categories may be two or more. The chi-square deals best with the type of data that falls in two or more discrete classes or categories. Such data is very common; some examples are attribute counts, ranked data, and the frequency of occurrence.

The χ^2 statistic is a measure of the degree of correlation. It is defined as:

$$c^2 = \frac{\Sigma(O-E)^2}{E} = \frac{\Sigma(Observed - Expected)^2}{Expected}, \text{ where}$$

O = observed number of cases or class frequency,

E = expected number of eases or class frequency,

and Σ directs the summation
over all classes or categories.

There are application restrictions, however. In general, it is not advisable to use the χ^2 test where any expected class frequency is less than five, although adjacent data may be grouped into a class to improve on class frequency.

The χ^2 distributions are continuous variables, so when the actual variable is not discrete, a correction must be made. Those values greater than expected should be reduced by 0.5, and those values that are less than expected should be increased by 0.5.

Chi-square values do not come from a single curve, but from a family of curves. A χ^2 graph (fig. 6.6) shows how each degree of freedom relates to one particular curve. This will be discussed in the next section. Table 6.3 is an abbreviated χ^2 table. Table A.4 in Appendix A is more complete.

Abbreviated Critical Values of the X² Distribution

Significant (Alpha ∝) Values

		.10	.05	.025	.01	.001
	1	2.71	3.84	5.02	6.64	10.83
	2	4.61	5.99	7.38	9.21	13.82
Degrees	4	7.78	9.49	11.14	13.28	18.47
of	6	10.65	12.59	14.45	16.81	22.46
Freedom	8	13.36	15.51	17.53	20.09	26.13
	10	15.99	18.31	20.48	23.21	29.59
	20	28.41	31.41	34.17	37.57	45.31

Table 6.3

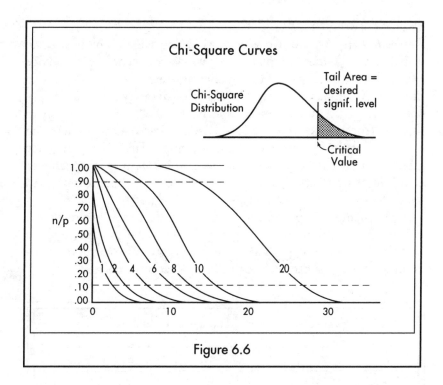

Chi-Square Curves

Figure 6.6

Chi-square: application example

The following example is based on the amount of maintenance required on two similar machines. The expected maintenance frequency is prorated because machine A receives more service than machine B, and it is more accurate to proportion the expected maintenance. The expected frequency may be calculated from the marginal totals. For instance, the expected service for machine A would be calculated by multiplying the total maintenance inspections for machine A by the total amount of service and dividing it by the grand total of maintenance inspections. Using the figures in table 6.5, the χ^2 statistic may be calculated using the chi-square formula:

$$\chi^2 = \frac{\Sigma(O - E)^2}{E} = \frac{\Sigma(Observed - Expected)^2}{Expected}$$

$$\chi^2 = \frac{(116 - 134)^2}{134} + \frac{(85 - 67)^2}{67} + \frac{(162 - 144)^2}{144} + \frac{(54 - 72)^2}{72}$$

$$= 2.41 + 4.83 + 2.25 + 4.50 = 13.99.$$

Use of chi-square table

Tables 6.4 and 6.5 are two-way classification tables. The horizontal classification divides the data into two classes according to service needs. The vertical classification divides the two classes into two different but similar machines. Classification of this type is called a *two-way contingency* table, and it determines if one classification is independent of the other. The statistical device used to test the hypothesis is the χ^2 test. How far must sample data in a contingency table depart from the proportional pattern before it is reasonable to conclude that, in the universe as a whole, the classifications are dependent?

As mentioned previously, the χ^2 statistic is a measure of the degree of correlation. Analysts generally apply the χ^2 test to those problems to determine whether the frequency of a current event's occurrence is significantly different from what was expected.

Prorated Expected Service Calculations		
	MACHINE A	MACHINE B
Expected Service Required	$\dfrac{278 \times 201}{417} = 134$	$\dfrac{139 \times 201}{417} = 67$
Expected Maintenance Not Required	$\dfrac{278 \times 216}{417} = 144$	$\dfrac{139 \times 216}{417} = 72$

Table 6.4

Machine Maintenance Records			
	MACHINE A	MACHINE B	TOTAL AMOUNT OF MAINTENANCE
Service Required	OBS = 116 EXP = 134	OBS = 85 EXP = 67	201
Maintenance Not Required	OBS = 162 EXP = 144	OBS = 54 EXP = 72	216
Total Maintenance	278	139	417

OBS = Observed maintenance required from maintenance records.
EXP = Expected maintence required–prorated.

Table 6.5

There are a number of different sampling distributions for the test, one for each degree of freedom (df). The number of degrees of freedom reflects the number of observations that are free to vary after certain restrictions have been placed on the data. These restrictions are not arbitrary, but are inherent in the data's organization.

Degrees of freedom are the number of data points minus one, i.e., n - 1. Stick diagrams such as that in figure 6.7 may be used as graphical representations of degrees of freedom. The dots represent data points and the connecting lines represent the degrees of freedom.

Degrees of Freedom Schematic

	Data Points	Degrees of Freedom
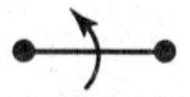	2	1
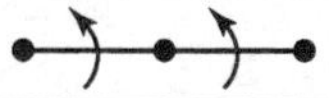	3	2
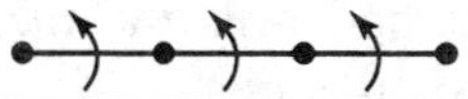	4	3

Figure 6.7

When used in a matrix, or a larger contingency table, the degrees of freedom are multiplied according to a simple formula: $DF = (r-1)(c-1)$. The row data points tabulate according to one variable and the columns according to another. This is used principally in the study of correlation between variables. Therefore, for the machine maintenance example, the degree of freedom is calculated as:

$DF = (2-1)(2-1) = 1$ degree of freedom.

When this is applied, the result is:

$$\chi^2 = 2.41 + 4.83 + 2.25 + 4.50 = 13.99$$
$$DF = (r-1)(c-1) = (2-1)(2-1) = 1$$

In table A.4 in Appendix A, for one degree of freedom, the χ^2 critical value at the .001 significance level is 10.83. Since the calculated χ^2 value of 13.99 is greater than 10.83, one can conclude that there is a highly significant difference in the amount of service required to maintain machines A and B.

The χ^2 table in Appendix A also gives values for probabilities greater than .10, namely .25, .50, .75, .90, .95, and .99. The last two values (.95 and .99) are those that fit better than they should. When data fits significantly better than it should, it is reasonable to suspect that the data is being "fudged."

An example of data fitting better than it should can be drawn from a coin toss. Suppose a coin is tossed 1000 times. If the result were exactly 500 "heads" and 500 "tails" on each trial, this would be remarkable. The expected results would be observed values of frequency corresponding to values of χ^2 scattered over the range of .95 to .05 with 1 out of 40 corresponding to the upper limit and one 1 of 40 corresponding to the lower limit. The value of χ^2 most frequently would be .50. Close fits should occur within the frequency of the χ^2 test. If it occurred more frequently, then "fudged" data is suspected.

One-sample analysis

A *one-sample analysis* procedure estimates and tests the mean and variance of a single random sample or two paired samples. The data may be used for four types of calculations:

- sample statistics;
- a confidence interval for the mean;
- a confidence interval for the variance; and
- a t-test for a hypothesis concerning the population mean.

A one-sample analysis usually involves the first sample in an investigation. It is very important that as much correct information as possible come from the sample because it is often the reference point for future analysis.

The sample should be large enough to represent the process distribution. Use table 5.1 to determine the number of measurements required to establish the variability with some precision. An examination of the binomial distribution (fig. 6.3) will provide a guide for the minimum sample size.

To ensure that the sample represents the population, randomizing the data selection is essential (see Chap. 5 for the procedure).

The shape of the distribution can be obtained from a probability plot using sample data (see Chap. 5).

Summary statistics (fig. 6.8) from a computer printout are indicative of the amount of statistical information that can be obtained from one sample. For example, the essential statistical information that can be obtained from a sample of 10 is:

DATA SET = 21, 40, 21, 21, 21, 31, 21, 40, 40, 43;

RANGE = High - Low = 43 - 21 = 22;

$$\text{X BAR} = \frac{\Sigma X}{n} =$$

$$\frac{(21 + 40 + 21 + 21 + 21 + 31 + 21 + 40 + 40 + 43)}{10};$$

$$\text{MEAN} = \overline{X} \text{ or } \overline{Y} = \frac{\Sigma X}{n} \text{ or } \frac{\Sigma Y}{n} = \frac{299}{10} = 29.9;$$

MEDIAN = middle measurement =

43, 40, 40, 40, 31, 21, 21, 21, 21, 21

$$\text{MODE} = \frac{31 + 21}{2} = 26$$

$$\text{VARIANCE} = s^2 = \frac{\Sigma(X_i - \overline{X})^2}{n-1} = \frac{\Sigma x^2 - (\frac{\Sigma x}{n})^2}{n-1}$$

$$= \frac{21^2 + 40^2 + 21^2 + ...43^2 - (\frac{299}{10})^2}{10-1}$$

$$= \frac{9815 - 8940}{9} + \frac{879}{9} = 97.2$$

$$\text{STANDARD DEVIATION} = s = \sqrt{s^2} = \sqrt{97.2} = 9.86$$

The confidence intervals for the mean and variance are computed using the t and chi-square distributions, respectively, with the assumption that the data comes from a normal distribution.

In addition to the above statistics, the following graphical techniques are applicable to the one-sample analysis:
- frequency distribution;
- check sheet;
- histogram;

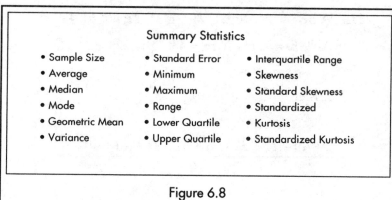

Figure 6.8

• stem-and-leaf histogram;
• hanging histobar; and
• probability plot.

In the one-sample analysis in table 6.6, a computer calculated the average, variance, standard deviation, and the median from the 154 observations.

The confidence interval for the mean was calculated for 95 percent of the expected values based on the t-distribution and a sample size

One-Sample Analysis Results

Sample Statistics:	Number of Observations	154
	Average	28.7935
	Variance	54.42325
	Standard Deviation	7.37721
	Median	28.9

Confidence Interval for Mean: 95 percent
                         Sample 1   27.62   29.97
                         Degrees of Freedom   153

Confidence Interval for Variance: Sample 1   0 percent

Hypothesis Test for H_0:   Mean   = 0
                    Computed t-Statistic = 48.4354
                    Significance Level   = 0
                    at Alpha   = 0.05
                    so reject H_0

Table 6.6

greater than 120 (see table A.3 in Appendix A). The level of significance-alpha 0.05-is for a two-tailed test (fig. 6.9) or .025 for a one-tailed test.

$$\frac{\bar{x} \pm ts}{\sqrt{n}} = \frac{28.7935 \pm 1.96(7.3772)}{\sqrt{154}} = 28.7935 \pm 1.165$$
$$= 27.62 \text{ and } 29.97.$$

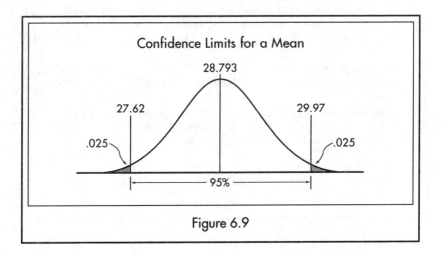

Figure 6.9

Two-sample analysis

The two-sample analysis procedure is used to estimate, test, and compare the means and variances of two samples. Like the one-sample, it can be analyzed by with numerical procedures and graphics. The data may be used for five types of calculations:

- sample statistics (*average, variance, standard deviation,* and *median*);
- a confidence interval for the difference between the two means, assuming equal variances;
- a confidence interval for the difference between the two means not assuming equal variances;
- a confidence interval for the ratio of the two variances; and
- a t-test for hypotheses concerning the differences between the two means.

A two-sample analysis from a computer analysis with statistical results is shown in table 6.7, where the sample statistics are displayed for each sample individually as well as for pooled samples. Also shown are 95 percent confidence intervals for the differences between two means, assuming equal and unequal variances.

Two-Sample Analysis Results

		Sample 1	Sample 2	Pooled
Sample Statistics:	Number of Obs.	106	106	212
	Average	3.83075	1.08208	2.45642
	Variance	2.97865	0.165104	1.57188
	Std. Deviation	1.72588	0.40633	1.25375
	Median	3.3	1	2

Confidence Interval for Diff. in Means: 95 percent
(Equal Vars.) Sample 1 - Sample 2   -3.08825   -2.40911   210 DF
(Unequal Vars.) Sample 1 - Sample 2   -3.08983   -2.40753   116.6 DF

Confidence Interval for Ratio of Variance: 0 percent
Sample 1 + Sample 2   0.0377302   0.0814307   105 DF

Hypothesis Test for H_0: Diff = 0   Computed t-Statistic = -15.9607
                       vs Alt: NE   Significance level = 0
                       at Alpha = 0.05   so Reject H_0

Table 6.7

Hypothesis testing

When comparing samples, a real difference must be identified, and this decision must be made despite uncertainty. It is assumed that there is no difference, and this is called the null hypothesis, denoted as H_0:

$$H_0: X_1 = X_2 \text{ (i.e., the means are equal), and}$$

$$H_0: S_1^2 = S_1^2 \text{ (i.e., the variances are equal).}$$

So, if $H_0: X_1 = X_2$ and $0_1^2 = 0_1^2$, the samples are identical.

Null hypothesis alternatives are not equal (NE), less than (LT), or greater than (GT). The NE (not equal) hypothesis is applied to a two-sided test, LT (less than) is for a one-sided test, and GT (greater than) for a one-sided test.

Making a wrong decision in testing a hypothesis is an important consideration. For example, if it is assumed that there is no difference ($H_0 = 0$) when one does exist, an error of the first kind would be made (see figs. 6.10 and 5.31). The probability of making this error is referred

Hypothesis Testing		
Condition	Hypothesis True	Hypothesis False
Commitment	Decision	
Accept True Hypothesis	Acceptable	Beta Type II Error
Reject False Hypothesis	Alpha Type I Error	Acceptable

Figure 6.10

to as *alpha* (α) *.05 or alpha .01*, which means that a wrong decision could be made 5 percent or 1 percent of the time, respectively. When making this test in a statistical analysis, the alpha percentage must be specified in advance, i.e., what risk is one willing to take that a wrong decision will be made?

F Ratio Data Example		
	Machine appears to be rough	Normal condition of Machine
Average	$\bar{X}_1 = .016$	$\bar{X}_2 = .015$
Variance	$s_1{}^2 = 41.60$	$s_2{}^2 = 18.1$
Degrees of Freedom	$n_1 = 10 - 1 = 9$	$n_2 = 60 - 1 = 59$

Table 6.8

The computed t-statistic is used to test whether the two means could be regarded as coming from the same population. Sizes for both samples must be considered to calculate the t-statistic. With the data from table 6.8, the t-statistic is computed as follows:

$$t = \frac{\overline{X}_1 - \overline{X}_2}{\sigma}\sqrt{\frac{n_1(n_2)}{n_1+n_2}} = \frac{3.831-1.082}{1.253}\sqrt{\frac{(106)(106)}{106+106}} = 15.96$$

for $(n_1 + n_2 - 2) = (106 + 106 - 2) = 210$ *degrees of freedom*.

Consulting the t-table (see table A.3 in Appendix A) for 210 degrees of freedom at the alpha 0.05 critical level, it is found that the t must be less than 1.96. Since 15.96 is much greater than 1.96, there is a significant difference in the distributions, and the null hypothesis must be rejected.

Comparing variances

Suppose that the data on a control chart shows what appears to be a rough condition on the R-chart, but the X-chart is running in control. If this is truly the case and the range (or variance) is deteriorating, stopping the machine for maintenance would be a major consideration. The assumption for statistical testing is, however, that no difference has taken place.

The first thing to determine is the initial process variation from the control chart. For example, if the R data average was 15 and this data came from samples of 5, the initial sample variance would be:

$$s = \frac{\overline{R}}{d_2} = \left(\frac{15}{2.326}\right)^2 = 6.45, \text{ where}$$

d_2 for samples of 5 is 2.326.

If these samples of 5 were taken from a control chart with 12 data points, the total size of the sample would be 5 x 12 or 60. The degrees of freedom (n_1) would be 60 - 1 = 59.

It is not advisable to run a machine continuously in what appears to be a rough condition. Instead, someone should stop the machine and take a random sample of 10 parts and determine if a significant change has taken place. The average and variance should be calculated for these 10 parts with the results in table 6.8.

The averages of the machine in the rough condition appear to be about the same as the normal condition of the machine, but the variance appears to have worsened. Can a decision be made on the basis of ten parts? Since the averages are the same, it could be assumed that no change has taken place.

To obtain the F-ratio, calculate the larger variance to the smaller:

$$\text{F-ratio} = \frac{s_1^{\,2}}{s_2^{\,2}} = \frac{41.60}{18.1} = 2.30$$

Associated with the variance ratio are two sets of degrees of freedom, n_2 for the smaller variance and n_1 for the larger variance. In most statistics books, there are F-ratio tables (e.g., table A.5 in Appendix A) with a wide set of values for nl and n_2 degrees of freedom for .10, .05, .01, and .001 levels of significance (see table 6.10).

If the value for an F-ratio is greater than that given in the table of critical values for the F-distribution, then the result is more significant than that level of significance.

Thus, for degrees of freedom $n_2 = 59$ and $n_1 = 9$ for the .25, .10, .05, .025, .01 and the .001 probability levels of significance, the F-ratio values from these tables are found in table 6.9.

The F-ratio from table 6.9 was 2.297. For the 59 degrees of freedom, the next higher value in the table is 60, and the 2.297 lies between the .05 and the .025 level of significance, or at about .02. This means that these results occur 2 percent of the time by chance or 2 times out of 100. The null hypothesis supposed that no difference existed; however, the data has shown this to be incorrect 98 percent of the time. Thus the null hypothesis is rejected, and it may be assumed that a change has most probably occurred and that the machine requires maintenance.

These F-ratio values are now being analyzed by computers, so it is important to understand how to interpret these statistics.

Abridged Table of Variance Ratio
F Distribution Points for $n_2 = 9$ DF

n_1 DF	.25	.10	.05	.025	.01	.001
10	1.56	2.35	3.02	3.78	4.94	8.96
20	1.41	1.96	2.39	2.84	3.46	5.24
30	1.36	1.85	2.21	2.57	3.07	4.39
40	1.34	1.79	2.12	2.45	2.89	4.02
60	1.31	1.74	2.04	2.33	2.72	3.69
120	1.29	1.68	1.96	2.22	2.56	3.38
∞	1.27	1.63	1.88	2.11	2.41	3.10

Table 6.9

Two-variable samples

When comparing two sets of data, it is useful to show an association or a correlation between them. Comparing one charted variable to another may yield a graph with points that lie along a straight line, curved line, or a scatter diagram, depending on the relationship. Correlation is a refinement of the scatter plots discussed earlier (see Chap. 4). The purpose of a correlation analysis is to determine the relationship between two variables: the independent variable (X) and a dependent variable (Y). When studying a functional relationship (cause and effect), where the response of Y is a function of the predicting variable X, the true relationship can be described by a line of best fit. This line is known as the *line of least squares* or *line of regression* (fig. 6.11).

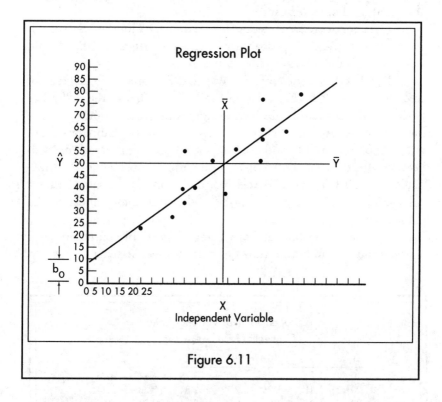

Figure 6.11

When only two characteristics are involved, the natural first step in handling the experimental results is to plot the points on graph paper. The independent variable X is plotted on the horizontal scale and the dependent variable Y, on the vertical scale.

A pictorial indication of the probable form and sharpness of the relationship, if any, is indispensable. In fact, there is no substitute for a plot of the data when it comes to indicating general spread and shape of the results. Sometimes it even prevents needless computing. The charted data will show whether a hypothetical linear relationship is borne out; if not, a theoretical basis for fitting a curve of higher degree must be found. During a search for an empirical association of two characteristics, a glance at the plot may reveal whether such association is likely or whether the scatter of points lacks a pattern. In the latter case, the data should be tested as outlined in Chapter 4.

If the plot reveals no obvious difficulties, and the relationship appears to be linear, then the line should be fitted to the data using the equation for a straight line:

$$Y = b_0 + b_1 X \text{ or } X = b_0 + b_1 Y.$$

The definition of the best line is that which deviates the least from the line of measurements of Y; the independent variable X is assumed to be free of error.

There is no guarantee that two variables in a complex system will have a straight line relationship. A straight-line analysis is most appropriate when the data plotted on a graph resembles an elliptical cloud and it is not known whether the cloud is caused by a genuine relationship or a sampling error. Alternatively, there might be little doubt about the reality of the existence of the relationship, but the best estimate of the relationship is desirable.

If the data falls on a smooth curve with a complex shape, then the straight-line techniques are not applicable. If there is a prior reason, from the theory of the process, to suspect that the relationships may have other forms, then a computer program or a probability plot should be used to fit the data.

The following are four regression models that can be fitted by computers:

$$\text{linear: } Y = a + bX;$$

$$\text{multiple: } Y = aX^b$$

$$\text{exponential: } Y = e^{a+bX}$$

$$\text{reciprocal: } \frac{1}{Y} = a + bX.$$

In the multiple and exponential models, linearity is achieved by logarithmic transformations. Model parameters are then estimated. In the reciprocal model, the reciprocal of the dependent variable is used.

To test for the significance of an apparently linear relationship, the *correlation coefficient* (r) is calculated:

$$r = \frac{\Sigma(X - \bar{X})(Y - \bar{Y})}{\sqrt{\Sigma(X - \bar{X})^2 \Sigma(Y - \bar{Y})^2}} \text{ , where}$$

Σ denotes summation over all pairs of observations.

If the relationship between the data can be represented by a straight line, then $r = \pm 1$. The 1 is positive if the straight line has a positive slope and negative if the straight line has a negative slope. If there is no relationship at all between variables, then $r = 0$. Even when there is no relationship, there is still a small value of r due to the variability and randomness of the sampling. This is the residual variance or residual error. This residual error includes experimental error and assignable sources of variation not taken into account by the model equation.

Like any other statistical test, the sample size with its associated degrees of freedom (df = n - 2), is required to determine the correlation coefficient (r). Its significance can be determined by consulting a table of correlation coefficients (r) for the appropriate degrees of freedom (see table A.6 in Appendix A).

Regression analysis

Regression analysis is similar to the analysis of variance. It is particularly pertinent when the factor levels are continuous and when there is more emphasis on the model equation than the hypothesis tests.

Table of Regression Analysis of Variance

Source	Sum of the Squares		DF	Variance Estimate
Regression	$r^2 \Sigma(Y - \bar{Y})^2$	= 2647.8	1	2647.8
Residual	$(1 - r^2)[\Sigma(Y - \bar{Y})^2]$	= 873.2	13	67.2
Total	$\Sigma(Y - \bar{Y})^2$	= 3521.0	14	

Table 6.10

The data in table 6.10 is used to evaluate the ability of early life test results to predict future life test data. Early life test results are the independent variables - X, and final life test results are the dependent variables - Y. The table shows the life test values for fifteen samples, together with the computations of the sum of squares and products. These are necessary for the computation of the regression coefficient and the correlation coefficient.

In the table, the independent variable X represents the early test data after 5,000 cycles and the dependent variable Y is the test data after 45,000 cycles. The variables have been transformed, however, to lessen the arithmetic. Plotting Y against X (fig. 6.13) produces a graph with a wide scatter but, apparently, a decided tendency for higher initial results to give higher final results. Is this relationship significant based on fifteen pairs of observed data?

$$s^2x = \Sigma(X - \overline{X})^2 = \Sigma(X^2) - \frac{(\Sigma X)^2}{n}$$

$$= (5^2 + 8^2 + 9^2 + \ldots + 20^2)^2 - \frac{(5 + 8 + 9 + \cdots + 20)^2}{15}$$

$$= 2689 - \frac{(191)^2}{15} = 2689 - 2432 = 257.$$

$$s^2y = \Sigma(Y - \overline{Y})^2 = \Sigma(Y)^2 - \frac{(\Sigma Y)^2}{n}$$

$$= (20^2 + 30^2 + 40^2 + \ldots + 75^2)^2 - \frac{(20 + 30 + 40 + \ldots + 75)^2}{15}$$

$$= 40921 - \frac{(749)^2}{15}$$

$$= 40921 - 34700$$

$$= 3521$$

$$s^2xy = \Sigma(X - \overline{X})(Y - \overline{Y}) = \Sigma(XY) - \frac{\Sigma(X)\Sigma(Y)^2}{n}$$

$$= 10362 - \left[\frac{(191)(749)}{15}\right]$$

$$= 10362 - 9537$$

$$= 825$$

The correlation coefficient (r) is then defined as:

$$r = \frac{\Sigma(X - \overline{X})(Y - \overline{Y})}{\sqrt{\Sigma(X - X)^2 \Sigma(Y - \overline{Y})^2}} = \frac{824.7}{\sqrt{257(3521)}}$$

$$= \frac{824.7}{951.3} = .8669$$

$$r^2 = .752$$

The degrees of freedom with n = 15 are:

$$n - 1 = 15 - 1 = 14.$$

The table of the correlation coefficients (table A.6 in Appendix A) for fourteen degrees of freedom shows that the probability of getting such a value for (r) in the absence of correlation is greater than .001. In other words, the chance of again attaining the same values for r is 1 in 1000. The evidence for correlation is therefore very strong.

The equation for the regression line of Y on X is:

$$\hat{Y} = b_0 + b_1X = \text{sample equation for a straight line, and}$$
$$\hat{Y} = b_0 + b_1(X - \overline{X}) = \text{equation for X variables, where}$$

$\hat{Y}$ = predicted value of Y;

X = predictor value of X;

b_0 = the intercept value of the regression line; and

b_1 = the slope of the regression line.

All of the observations (X, Y) fall on the line except for the residual error, which is the difference between the observed results and the predicted value. Residual error includes experimental error and assignable sources of variation not taken into account by the model. b_0 and b_1 are the constants that minimize the sum of the squared deviations between the observed response (Y) and the predicted response (Y).

To predict values of (Y) from known values of (X), the following calculations are made from the data in table 6.11:

$$b_0 = \frac{\Sigma Y}{n} = \frac{749}{15} = 49.9 \text{ (regression intercept)},$$

$$b_1 = \frac{\Sigma(X - \overline{X})(Y - \overline{Y})}{\sqrt{\Sigma(X - X)^2 \Sigma(Y - \overline{Y})^2}} = \frac{824.7}{257} = 3.20 \text{ (regression slope)}$$

$$\overline{x} = \frac{\Sigma X}{n} = \frac{191}{15} = 12.73$$

The formula for the most probable value of Y corresponding to a given value of (X) is then:

$$\hat{y} = b_0 + b_1(X - \overline{X})$$
$$= 49.9 + 3.20(X - 12.73)$$
$$= 49.9 + 3.20X - 40.7$$
$$= 9.2 + 3.2X$$

This is the best estimate of Y from a known value of X.

Regression analysis of variance
As previously calculated, the total sum of squares is:

$$\Sigma(y - \overline{y})^2 = \Sigma y^2 - \frac{(\Sigma Y)^2}{n} = 40921 - \frac{(749)^2}{15} = 3521.$$

The portion of the total sum of squares accounted for by regression is:

$$(r^2)\left[\Sigma(Y - Y)^2\right] = (.752)(3521) = 2647.8$$

			Regression Analysis Data		
n	X	Y	X^2	Y^2	XY
1	5	20	25	400	100
2	8	30	64	900	240
3	9	40	81	1600	360
4	9	35	81	1225	315
5	9	55	81	3025	495
6	10	42	100	1764	420
7	12	50	144	2500	600
8	13	38	169	1444	494
9	14	55	196	3025	770
10	16	50	256	2500	800
11	16	62	256	3844	992
12	16	75	256	5625	1200
13	16	60	256	3600	960
14	18	62	324	3844	1116
15	20	75	400	5625	1500
Totals	191	749	2689	40921	10362

Table 6.11

The portion of the total sum of squares not accounted for by regression is the total sum of the squares minus the regression sum of squares:

$$(1 - r^2)\left[\Sigma(Y - Y)^2\right] = (1 - .752)(3521)$$

$$= 873.2, \text{ or}$$

$$3521 - 2647.8 = 8732.2$$

A regression takes 1 degree of freedom; therefore the degree of freedom for the residual is 14 - 1 = 13. These calculations are shown in table 6.12.

Residual variance about the regression line

The null hypothesis assumes that the regression coefficient, r, is 0. The variance estimates, then, are estimates of the residual (i.e. non-regression) variation, and the F-ratio is:

$$\frac{Regression\ Estimate}{Residual\ Estimate} = \frac{2647.8}{67.2} = 39.4$$

with 1 and 13 degrees of freedom.

Tables of F-distribution critical values show that 39.4 exceeds the .001 significance level, 17.8. The regression coefficient is therefore significant. This figure demonstrates the usefulness of the regression line equation for predicting values of Y for given values of X.

Suppose confidence limits are drawn on either side of the regression line, which would contain 95 percent of all points. According to the t table of critical values, the value for t when n - 2 = 13 degrees of freedom and the level of significance is .05, the residual variance about the regression line is:

$$t_{.05} = 2.16$$

$$s = \sqrt{67.2} = 8.2$$

$$ts_r = (2.16)(8.2) = 16.4$$

Two lines parallel to the regression line are drawn in; one displaced 16.4 units of Y downward and the other, 16.4 units of Y upwards. When the regression line is used to predict values of Y from a known value of X, these confidence limits provide a guide for 95% correctness.

Multivariate methods

Multivariate correlation involves two or more independent variables. Multivariate data samples can be analyzed by some of the same techniques used for a two-sample analysis. Table 6.12 is a summary of these methods. Usually examined by numerical techniques, they may also be studied on certain types of graphs. Since some of the procedures require more than an elementary knowledge of statistical methods, only some of the basic ideas will be presented here.

Multiple correlation analysis

A *multiple correlation analysis* uses a matrix of correlation coefficients for sets of observed values. It provides a preliminary view of the relationships among a number of variables.

The correlation coefficient values fall between -1 and +1. A positive correlation indicates that the variables vary in the same direction while negative correlation indicates a variance in the opposite direction (see fig. 4.6). Independent variables with no relationship have an expected correlation factor that approaches zero.

Three statistics are required for each pair of observations in the contingency table:

Multivariate Methods

Procedure	Number of Variables	Description
Correlation Analysis	2 or more	Analysis of Correlation Coefficients
Box-and-Whisker Plots	2 or more	Comparing Box-and-Whisker Plots
Probability Plots	2 or more	Comparing Probability Plots on the same sheet of probability paper
Multivariate Scatter Plots	3 or more	3-Dimensional Scatter Plots
Factorial Designed Experiments	2 or more	Analysis of Variance
Multiple Regression	2 or more	Analysis of Variance of Regression Equation

Table 6.12

• the correlation coefficient (r);
• the sample size; and
• the significance level.

Figure 6.12 is a *correlation matrix* created from variables from a data set called life testing of automobile axles. The files used were:
• life test results (end of life cycles);
• rockwell hardness of the axle (Rc);
• heat treat temperature (degrees c); and
• metallurgical variable (percent carbon).

Each correlation coefficient (r) is displayed in the contingency table with its corresponding sample size n (in parenthesis) and significance level. The significance level is based on the student's t-distribution. If that level is small (less than .05), then there is a significant correlation.

In the matrix, each variable is correlated with itself, resulting in a perfect correlation (r = 1.0000) and a very high significance level (.0000). The life test results and the rockwell hardness are negatively correlated (r = -.463).

The sample sizes for all coefficients were the same, i.e. (30), since the data came from a data file with all the possible combinations.

The contingency matrix determines the variables that are not significant and highlights those areas that need to be investigated further.

Multiple regression analysis of the significant variables is the next step in the analysis procedure.

	Hardness Rc	Heat Treat °C	Carbon %	Life Test Cycles (Y)
Contingency Matrix Life Testing of Automobile Axles				
Hardness Rc (X₁)	1.0000 (30) .0000	-.080 (30) 1.0000	.433 (30) .05	-.463 (30) .01
Heat Treat °C (X₂)	.080 (30) 1.0000	1.0000 (30) .0000	-.053 (30) 1.000	.352 (30) .05
Carbon % (X₃)	.433 (30) .05	-.053 (30) 1.000	1.0000 (30) .0000	-.087 (30) .94
Life Test Cycles (Y)	-.463 (30) .01	.352 (30) .05	-.087 (30) .94	1.0000 (30) .0000

Figure 6.12

Multiple box-and-whisker plots
If the observed values being examined can be subdivided into groups, then a multiple box-and-whisker plot is useful for examining each group. It can provide a great deal of information in simple-to-interpret graphs.

A multiple box-and-whisker plot is shown in figure 6.13. In this example, the same part was run on seven different machines, and there was an excessive variation in the assembled product using this part. To help identify the cause of this variation, a box-and-whisker plot was constructed from the data from each machine.

The source of variation became apparent when the four box-and-whisker plots were compared. The task then became one of minimizing the difference between machines.

Multiple probability plots
Multiple probabiltiy plots are ideal for analyzing and evaluating a designed experiment. They may be used to organize, summarize, and compare the various factors introduced into a designed experiment.

Each factor is automatically tested for normality, which allows factor distributions to be compared with the specification limits drawn on the plots. This is covered in detail in the next chapter.

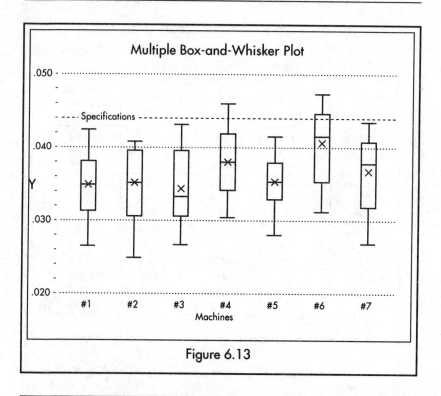

Figure 6.13

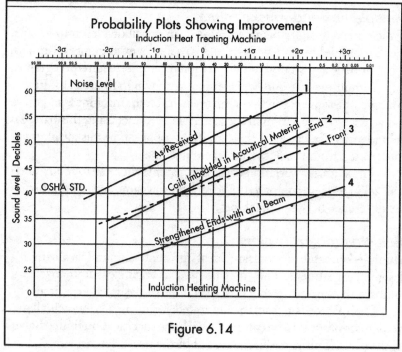

Figure 6.14

Such comparisons are useful reporting devices for presenting experiment results to managers that might not understand the mathematics of the analysis of variance, but can easily understand graphs. They are an important aid in the decision-making process.

When analyzing a designed experiment, it is best to plot each factor, along with the different levels of the factor, on a separate probability sheet. This allows factor comparisons without crowding.

Multiple probability plots may also be used to demonstrate improvements in a process. Figure 6.14 tracks the improvements made on an induction heat treating machine.

As received, the machine did not meet OSHA standards. The vendor provided insulation to the cabinets, but to no avail.

The plant engineers then provided more structural support to the machine to lessen vibrations. With this modification, the machine then met the OSHA standards. The probability plot was used in vendor negotiations regarding the cost of the improvements.

Conclusion

This chapter has discussed various methods for analyzing data using statistics. In some situations, a combination of statistical analysis and visual analysis works best. The conclusions drawn from these analyses can be used to prove a hypothesis. The next chapter discusses how to statistical and graphical analysis. That is step 4 in the plan for investigation.

Chapter 7

Process Improvement

"Leave the beaten path occasionally and delve into the woods; you will be sure to see something that you have never seen before."
                                                    -Alexander Graham Bell

"We have usually no knowledge that any one factor will exert its effects independently of all others that can be varied, or that its effects are particularly related to variations in these other factors. . . . if the investigator, in these circumstances, confines his attention to any single factor, we may infer either that he is the unfortunate victim of a doctrinaire theory as to how experimentation should proceed, or that time, material, or equipment at his disposal is too limited to allow him to give attention to more than one narrow aspect of his problem."
                                                    -R. A. Fisher

Process improvement results from reducing the variability of the process. This is easier said than done, because the process variability comes from many sources. How does one reduce this variability? First, grasp the idea that process variability comes from many sources: independent or dependent variables or interactions among these variables. This is the bad news. The good news is that an improvement in any of these variables will improve the overall process. Figure 7.1 is a mathematical equation that illustrates this point. A process is termed *capable* after its *unnatural* behavior problems have been eliminated. Further improvement comes from reducing the *natural* process variation.

Total Process Variation

$$\sigma^1 = \sqrt{\sigma_M^2 + \sigma_O^2 + \sigma_E^2 + \sigma_C^2 + \sigma_I^2 + \sigma^2 \text{ Residual}}$$

σ^1 = Total Process Standard Deviation

σ_M^2 = Variations From:
Machines
Methods
Materials

σ_O^2 = Variations From:
Operators
Shifts

σ_E^2 = Variations from Environment:
Temperature
Humidity
Time
Economics
Electric Power

σ_C^2 = Variations from Controlling Factors

σ_I^2 = Variations from Interactions

σ^2 Residual = Residual Variations
Measurement Errors
Unknown Factors

Figure 7.1

Control charts will indicate the more obvious changes in unnatural patterns *(assignable causes)*. The X-bar portion of the chart, or the *process average*, will indicate the direction of the change needed to improve the process. After that, any new improvements to the process must come from improving the basic process variability. This improvement will be identified on the range chart or by a reduction in slope *(standard deviation)* on a probability plot.

When a process improvement study is made, it should be recognized that the total process variance is the sum of the system variances, e.g., example, equipment, operator, material, and environmental variances, as well as interactions. Often, these unknown interactions prevent process improvement and may result in the rejection of otherwise acceptable variation as a cause effect. Each source of variability has its own unique effects as well as the possibility of interacting with other variables. Process improvement comes about by identifying the contributing variables finding ways to reduce them.

Contributing variables are *independent variables* and, together with any possible *interactions*, they result in an increase in the *dependent response variables* (fig. 7.2).

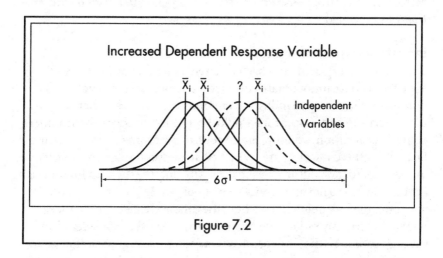

Figure 7.2

Thus, any reduction in any independent variable will decrease the total response variable. This means that a process can be improved upon by decreasing any contributing variable, as long as it does not have a negative effect on another, interactive variable.

A *designed experiment* and *an analysis of variance* are used to determine how much the independent variables and their interactions contribute to the total response.

Classical experimentation

The traditional method of industrial experimentation is to try one thing and if that does not work, try something else, and so forth. This procedure is inefficient. There is no way to distinguish the interrelated factors, variations in the environment, and experimental error. In addition, the controlled variable may be interacting with an uncontrolled variable or a variable being held constant.

Research engineers are often in the fortunate position of having all their independent variables under complete control. They know the sources of their materials and their measuring instruments have been calibrated. Their environment can be carefully controlled, and they can make measurements on the dependent variables. Thus, with a well-equipped laboratory, it is easy to hold all but one variable constant and then determine the effect(s) the one controlled variable will have. However, even with time, money, material, and equipment at his or her disposal, the research engineer is able to focus only on a very narrow aspect of the problem. Any interactions of the independent variables on the dependent variable cannot be determined. Also, there is no guarantee that the one factor under control will exert its effects independently of all the others that can be varied.

Industrial experimentation

The industrial experimenter has a different and more difficult problem. He or she often cannot obtain complete control of all the variables. The number of factors in an industrial environment is so vast that an army of supervisors and inspectors would be needed to keep them under control. In addition, attaining good control over some factors may also be a difficult technical matter. Further, it is essential for an industrial investigator to carry out an investigation with only the slightest hindrance (or none at all) to normal production. Thus, while it might be possible to control the independent variables, the time and care required for each lot would seriously reduce the plant's output. Another consideration is varying some of the independent variables over a wide range to determine the effects. This, however, might lead to a large amount of out-of-specification (scrap) material.

Designed experiments

An orderly procedure that results in the most information with a minimum of changes (variables) constitutes a designed experiment. A properly designed experiment allows relatively simple statistical interpretation of the results, which may not be possible otherwise.

Design of experiments (or experimental designs) is the efficient arrangement of experimental programs for studying the effects of controlled variables. This arrangement includes randomization (see Chap. 4). Each variable under investigation is a factor and each variation of the factor is called a level. As defined in Chapter 1, factors may be quantities, such as temperature or revolutions per minute, that can be varied along a continuous scale. Or, they may be qualitative: different machines, operators, and compositions of materials.

A *factorial designed experiment* can control several factors and investigate their effects at each of two or more *levels*. If two levels of each factor are involved, the experimental plan consists of taking an *observation* at each of the 2^n possible combinations. Figures 7.3 and 7.4 and Appendix C outline some designs of experiments.

In addition to the factors, which are varied in a controlled fashion, the experimenter should be aware of certain background variables that might affect the outcome of the experiment. For one reason or another, these background variables (such as ambient temperature) will not or cannot be included as factors in the experiment. However, it is often possible to plan the experiment so that:
- possible effects from background variables do not affect information about the factors of primary interest; and
- some information about these effects can be obtained.

There also may be undetected variables that may have an effect on the outcome of the experiment. The effect of these variables may be given an opportunity to balance out by the introduction of randomization into the experimental pattern.

Randomization

Randomization is necessary in order to eliminate bias from the experiment. Experimental variables that are not specifically controlled as factors or "blocked out" by planned grouping should be randomized by some method of randomization. This may be achieved by assigning consecutive numbers to each factor. Then, by using a table of random numbers, determine the random order in which the factor portion of the experiment should be run (see Chap. 4).

Replication

To evaluate the effects of experimental factors, a measure of precision (experimental error) must be available. In some industrial experimentation, records may be available on a relatively stable measurement process. For example, the R-data from a control chart would be an appropriate measure.

In some experiments, this measure should come from the experiment itself, since no other source would provide an appropriate measure. In this situation, replication provides the measure. It allows uncontrolled factors to balance out, which also helps decrease bias. Replication can also bring to light gross measurement errors.

Randomized blocks

One of the simplest concepts in design of experiments is the randomized block. A block is a homogeneous unit, usually selected to allow for assignable causes, in addition to those introduced as factors in the experiment (principal factors). It is usually assumed that the block factors do not interact with the principal factors.

Suppose it is necessary to compare four treatments–a combination of each factor's levels–assigned to an experimental unit. For example, four treatments may be four temperatures or four operators or four types of materials or four different machines of the same type or four spindles, and so forth. The materials to which the treatments are applied are in *batches* large enough to allow for four experiments on each batch.

Successive batches, or blocks, will differ in their quality. The operators just mentioned may be considered a block. They are more likely to be similar in training and experience than operators in another plant.

In the case of the machines, it is probable that the machines are from the same manufacturer and are the same models. If the treatments are A, B, C, and D, and the blocks are 1, 2, 3, and 4, then replicating the experiment four times produces the results shown in figure 7.3.

In the designed experiment illustrated in figure 7.3, each treatment is allocated once to each block. The treatments are further distributed in a planned random manner. When the average of the four As is compared with the average of the four Bs, the difference is completely independent of any block. At the same time, each block is receiving all four treatments. This is the concept of a *randomized block*: all treatments are carried out an equal number of times in each block.

Replication is achieved by repeating the experiment on further blocks. Several replications may also be placed in each block. This replication will give a better estimate for the residual error. This is important in determining the significance of the experiment's results, which will be discussed later in this chapter.

Randomized Block					Balanced Incomplete Block				
Block	Treatment				Batches	Treatment			
1	A	B	C	D	1	A	B	C	D
2	B	A	D	C	2	A	B	C	E
3	C	D	A	B	3	A	B	D	E
4	D	C	B	A	4	A	C	D	E
					5	B	C	D	E

Figure 7.3

Balanced incomplete blocks

The *balanced incomplete block* shown in figure 7.3 is an improved method where five treatments–A, B, C, D, and E–are replicated four times in five batches of four. Each treatment occurs once and only once in four of the five batches. A pair of treatments occurs together in three of the five batches:

- A and B–three times;
- A and C–three times;
- A and D–three times;
- A and E–three times;
- B and C–three times;
- B and E–three times;
- D and E–three times;
- C and D–three times;
- C and E–three times; and
- B and D–three times.

For example, B and D occur together in batches 1, 3, and 5. A comparison, or first order interaction, can be made for each treatment three times. Naturally, this is better than a single comparison.

This type of experimental design is called a balanced incomplete block. It is balanced because each treatment occurs exactly to the same extent. It is incomplete because no block contains the full number of treatments.

Factorial design of experiments

> "In the infinitely difficult act of thinking, nothing is more difficult than to separate what is known from what is not known—unless it be to understand that the separation must be made."
>
> -Bernard deVoto, *The Course of Empire*

The type of experiments discussed so far have had only one *independent variable* or *factor*. In most industrial work, there are a large number of independent variables that need to be investigated. A *factorial experiment* is an experiment wherein several factors are controlled and their effects on each of two or more *levels* are investigated. The experimental design consists of taking an observation at each one of all combinations formed by the different factor levels. Each is a different *treatment* combination. There also may be an *interaction* effect between any two factors in the range of interest. In the factorial experiment, each observation is used many times over but in a different manner each time. There is usually no knowledge that any one factor will exert its effect independently of all others that may vary, or that its effects are related to the variations in other factors.

The *factorial design of experiments* detects this type of effect and at the same time it provides the maximum amount of information about the process being tested for a given amount of labor.

The technique of factorial experimentation combined with a statistical technique for analyzing the results was first proposed in 1923 by Sir Ronald A. Fisher, a distinguished British statistician. Fisher developed it for agricultural experiments at the Rothamsted Experiment Station. With F. Yates, Fisher developed the statistical tables for biological, agricultural, and medical research used in the analysis of factorial experiments.

Fisher's approach to experimentation differs in two fundamental aspects from the classical, one-variable-at-a-time system. He stressed the importance of obtaining an accurate estimate of the error variation magnitude rather than its minimum amount. An accurate estimate is necessary to apply an exact test of significance, which will be discussed later in this chapter. Secondly, he stressed the advantages of including in the same experiment as many possible factors whose effects, *dependent variables*, are to be determined. The advantage of this, in addition to efficiency, is the determination of the extent to which the factors interact. This gives the experimenter a wider basis for any conclusions that may be reached.

Since the early development and applications were in agriculture, the terms used to describe the experimental designs were associated with that field. A *block* was a piece of ground, small enough to be fairly uniform in soil and topography, and thus expected to give results that would be more alike. A *plot* was an even smaller piece of ground and the basic unit of the design. A treatment actually was a treatment, such as the application of fertilizer. A *yield* was the quantity harvested and weighed or measured.

For process control, a block may be a group of results from a particular machine or particular operator or particular shift. Any planned, natural grouping serves to make one block more alike than results from different blocks. A *treatment* is the factor being investigated in a single factor experiment. In factorial experiments, where several variables are being investigated at the same time, a *treatment combination* would be the prescribed levels of the factors to be applied to an experimental unit. For example, for various machine speeds, a yield would be a measured result, such as output rate, and that would be the *dependent* variable.

The field of experimental statistics is quite extensive, and many good books on experimental design and analysis are available. Examples of some designs and their analyses are discussed in the sections that follow.

Latin square and Graeco-Latin square

In a *Latin square* and a *Graeco-Latin square*, the treatments are grouped into replicates in two different ways. Every row and every column of every square are complete replications. This eliminates all differences among rows and equally all differences among the columns from the errors. These squares provide more opportunity than randomized blocks to reduce the experimental error through skillful planning. The experiment should be arranged and conducted so that the difference among columns and rows represent major sources of variation.

The *Latin square* (fig. 7.4) is a design involving three factors, where the combination of any one of them with the versions of the other two appears once and only once. Versions of the third factor are shown by the Roman ("Latin") letters.

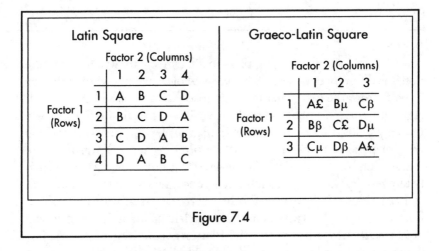

Figure 7.4

A design that involves four factors, where the combinations of any one of them with the versions of the other three appear once and only once, is called a *Graeco-Latin square* (fig. 7.4). Versions of factor 3 are shown by Latin letters and factor 4 by Greek letters.

Latin square example

An experiment used by Fisher to illustrate the principles of randomization and experimentation is an example of the practical application of probability plots to a Latin square (Fisher, 1932). It utilized the principles of randomization and replication on potato fertilization. The design of the experiment was a 6 x 6 Latin square (fig. 7.5). Every row and every column was a complete replication.

Arrangement and Yields of a 6x6 Latin Square

	Fertilizer Yields						Totals
	E 633	B 527	F 652	A 390	C 504	D 416	3122
	B 489	C 475	D 415	E 488	F 571	A 282	2720
Yield Data = Pounds of Potatoes	A 384	E 481	C 483	B 422	D 334	F 646	2750
	F 620	D 448	E 505	C 439	A 323	B 384	2719
	D 452	A 432	B 411	F 617	E 594	C 466	2972
	C 500	F 505	A 259	D 366	B 326	E 420	2376
Totals	3078	2868	2868	2722	2625	2614	16659

Figure 7.5

The object of arranging plots in a Latin square was to eliminate possible differences in fertility between whole rows and whole columns of plots from the experimental comparisons as they stood in the field. Agricultural experimenters saw the need for double elimination because in many fields, there was either a gradient of fertility across the whole area, or parallel strips of land with fertility higher or lower than the average. For particular fields, however, it was not known whether such heterogeneity was more pronounced in one direction or the other. Such soil variations may have been due in part to the past history of the field. For example, the land in which it had been laid up for drainage may have produced differences in the depth and present condition of the soil. Or, the soil may have been manured or cropped otherwise. Whatever the causes, the effects were sufficiently widespread to make apparent the importance of eliminating the major effects of soil heterogeneity, not only in one direction across the field but in the direction at right angles to it.

The 6 x 6 Latin square in figure 7.5 was designed as if nothing were known of the treatments applied, or as if they were merely causes disturbing the yields with an unknown variance. The six-column and six-row treatments were, therefore, not considered homogenous among themselves, but were subdivided into unitary elements of a very different agricultural importance.

Actually, the experimenters knew that the treatments D, E, and F differed from A, B, and C by including an additional nitrogenous dressing. In like manner, A, B, C, and D, E, and F differed among themselves by receiving, respectively, 0, 1, and 2 units of a phosphoric dressing. In table 7.1, the total yields of the six treatments are shown in relation to the material treatment received.

Design of Experiment					
	No Nitrogen		Nitrogen		Total
No Phosphate	A	2070	D	2431	4501
Single Phosphate	B	2559	E	3121	5680
Double Phosphate	C	2867	F	3611	6478
Total		7496		9163	16659

Table 7.1

Because the experiment was well designed, each treatment corresponded with a system of appropriate hypotheses relevant to the aims of the experiment. The effects due to nitrogen were highly

Analysis of Variance					
Component of Treatment	Sum of Squares	Degrees of Freedom	Mean Square	F Ratio	Signif.
N	77,191	1	77,191	21.2	< .001
P^1	162,855	1	162,855	44.7	< .001
P^2	2,016	1	2,016	.5	None
$N P_1$	6,112	1	6,112	1.7	.20
$N P_2$	5	1	5	-	None
Residual	109,208	30	3,640		
Total	357,387	35	10,211		

Table 7.2

significant, less than .001 (table 7.2). This conclusion referred directly to the nitrogen fertilizer comparison on which it was based and was entirely independent of the other conclusions (main effects) to be drawn regarding the effect of phosphate or of its interaction with nitrogen.

The thirty-six data points were checked for normality on probability paper (see Appendix H) and then the following procedure was used for this experiment:

1. The data values for each factor were put in order, from the smallest to the largest. The rank of 1 was assigned to the smallest observation, the rank 2 to the second smallest, and so forth; n was given to the largest observation as shown in tables 7.3, 7.4, 7.5, and 7.6:
    • n = 36 for all data yields from the Latin square (table 7.3);
    • n = 18 for the nitrogen fertilizer yields (table 7.4);
    • n = 12 for the phosphate fertilizer yields (table 7.5); and
    • n = 6 for the interacting plot yields for A, B, C, D, E, and F (table 7.6).

Data for Ranked Plotting Points and Observations
Ranked Data on Fertilizer Yields from Latin Square Plots

Rank	Plotting Points		Rank	Plotting Points		Rank	Plotting Points	
i	100(i-0.5)/n	Obser.	i	100(i-0.5)/n	Obser.	i	100(i-0.5)/n	Obser.
1	1.4	259	13	34.7	420	25	68.1	500
2	4.2	282	14	37.5	422	26	70.8	504
3	6.9	323	15	40.3	432	27	73.6	505
4	9.7	326	16	43.1	439	28	76.4	505
5	12.5	334	17	45.8	448	29	79.2	527
6	15.3	366	18	48.6	452	30	81.9	571
7	18.1	384	19	51.4	466	31	84.7	594
8	20.8	384	20	54.2	475	32	87.5	617
9	23.6	390	21	56.9	481	33	90.3	620
10	26.4	411	22	59.7	483	34	93.1	633
11	29.2	415	23	62.5	488	35	95.8	646
12	31.9	416	24	65.3	489	36	98.6	652

Table 7.3

Ranked Data on Nitrogen Fertilizer Yields from Latin Square Plots								
Rank	Plotting Points	No N^2 ABC	N^2 DEF		Rank	Plotting Points	No N^2 ABC	N^2 DEF
i	100(i-0.5)/n	Obser.	Obser.		i	100(i-0.5)/n	Obser.	Obser.
1	2.8	259	334		10	52.8	432	505
2	8.3	282	366		11	58.3	439	505
3	13.9	323	415		12	63.9	466	571
4	19.4	326	416		13	69.4	475	594
5	25.0	384	420		14	75.0	483	617
6	30.6	384	448		15	80.6	489	620
7	36.1	390	452		16	86.1	500	633
8	41.7	411	481		17	91.7	504	646
9	47.2	422	488		18	97.2	527	652

Table 7.4

2. Plotting positions for each observation were calculated from the formula:

$$Fi = \frac{100(i-5)}{n}, \text{ where}$$

$$i = 1 \ldots n.$$

The table in Appendix A, which is shown for n = 1 to n = 50, could also have been used.

3. On probability paper, the data scale on the y axis was labeled to span the experimental data. Each observation was plotted (fig. 7.6).

Ranked Data on Phosphate Fertilizer Yields from Latin Square Plots									
Rank	Plotting Points	No P AD	P^1 BE	P^2 CF	Rank	Plotting Points	No P AD	P^1 BE	P^2 CF
i	100(i-0.5)/n	Obser.	Obser.	Obser.	i	100(i-0.5)/n	Obser.	Obser.	Obser.
1	4.2	259	326	439	7	54.2	390	488	505
2	12.5	282	384	466	8	62.5	415	489	571
3	20.8	323	411	475	9	70.8	416	505	617
4	29.2	334	420	483	10	79.2	432	527	620
5	37.5	366	422	500	11	87.5	448	594	646
6	45.8	384	481	504	12	95.8	452	633	652

Table 7.5

Rank	Plotting Points	No P A	P1 B	P2 C	No P D	P1 E	P2 F
i	100(i-0.5)/n	Obser.	Obser.	Obser.	Obser.	Obser.	Obser.
1	8.3	259	326	439	334	420	505
2	25.0	282	384	466	366	481	571
3	41.7	323	411	475	415	488	617
4	58.3	384	422	483	416	505	620
5	75.0	390	489	500	448	594	646
6	91.7	432	527	504	452	633	652

Ranked Data on Phosphate Fertilizer Yields from Latin Square Plots

Table 7.6

4. A line of best fit was drawn through the plotted data.

5. Experimenters then studied the shape of the distribution plots in relationship to the dependent variable.

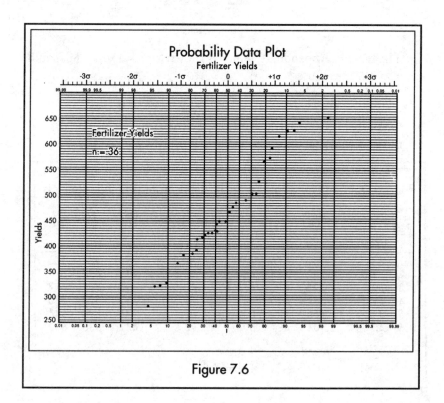

Figure 7.6

In the probability plot in figure 7.7, seventy-five percent of the data appeared quite normal, but a perturbation showed in a favorable direction for the other 25 percent. A further analysis was indicated.

The probability plot for the nitrogen factors (fig. 7.8) showed a significant main effect for the nitrogen fertilizer plots, D, E, and F, entirely independent of other conclusions. This is supported by the analysis of variance (see table 7.2). A perturbation, however, is indicated as an interaction with phosphate in F.

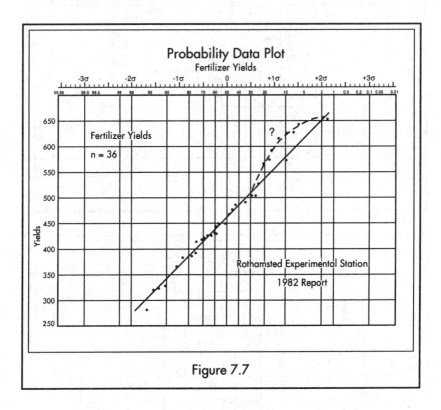

Figure 7.7

The primary effects of a single application of phosphates, B and E (fig. 7.9), were demonstrated with unquestioned significance and the magnitude of the return on yields was evaluated with fair accuracy.

The yield from a double application of phosphates, C and F (fig. 7.9), was also significant and in the right direction. These plots indicate that the effects would be definitely higher with higher phosphate dressings. However, the perturbation in the case of nitrogen-double phosphate, F, was still evident, showing a positive effect that agreed with agricultural experience. Interestingly, table 7.2 shows no significance.

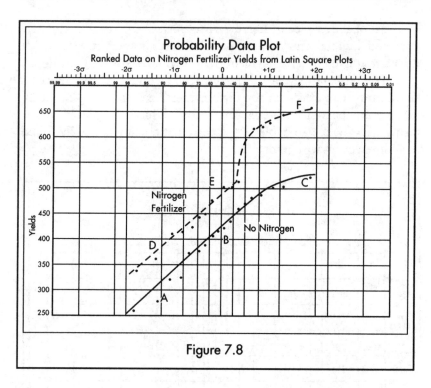

Figure 7.8

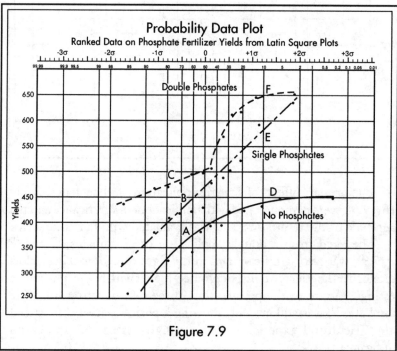

Figure 7.9

The interactive data from plot yields A, B, C, E, and F (fig. 7.10) provide more information on the interaction of nitrogen with single and double phosphate applications. The perturbation, F, with its double application of phosphate, is clearly in evidence. This plot shows that the double application of phosphate, although still charted in the right direction, is not as great as might be expected. This indicates that either not enough phosphate was applied or the double application was having a negative effect. Either way, further experimentation was indicated.

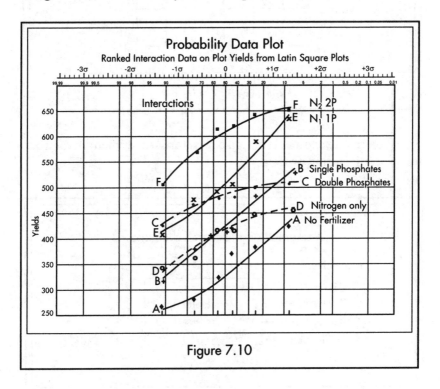

Figure 7.10

Factorial experiment

In a factorial experiment, all possible factors (treatment combinations) are formed from two or more factors, each being studied at two or more versions (levels) so that interactions (different effects) and main effects can be estimated.

Symbolically, a factorial experiment is the product of a number of versions (levels) of each factor. For example, 3 levels of factor A, 2 versions of factor B, and 4 versions of factor C is a 3 x 2 x 4 factorial experiment (fig. 7.11). The product of these numbers (24) indicates the number of treatments.

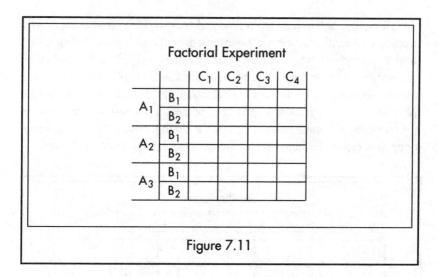

Figure 7.11

Figure 7.12 shows the design of a factorial experiment framework for a manufacturing process.

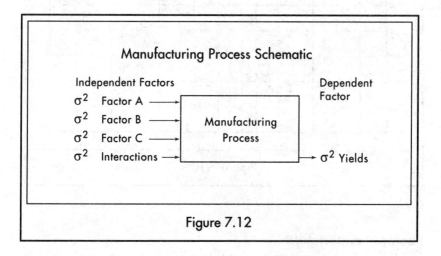

Figure 7.12

Fractional factorial experiments

For a complicated process, a complete factorial experiment that includes all possible combinations of factors may require many tests, even when only two levels of each factor are being investigated. In such cases, the complete factorial experiment may not only overtax the available facilities but also time and manpower. This happens to be the case in many manufacturing operations.

When a large number of treatment combinations result from a large number of factors, it is often impractical to test all the combinations in one experiment. In such cases, analysts may resort to a fractional, or partial, replication. Fractional factorial designs are often used in screening tests to determine which factor or factors are effective, or as part of a sequential series of tests or operations. The usefulness of these designs stems from the fact that, in general, higher order interactions are not likely to occur.

In other situations, it may not be practical to plan the entire experimental program in advance. In these cases, smaller experiments that need fewer tests serve as a guide for future work. There are, however, risks such as bad values or misjudgments. Fractional factorial experimental designs are sensitive to bad values. Bad values affect all other data.

The fraction is a carefully prescribed subset of all possible combinations; its analysis is relatively straightforward and uses analysis of variance techniques. The use of a fractional factorial experiment does not preclude the possibility of a later completion of the full experiment.

Fractional factorial experiments obviously are not as informative as the full factorial. Economy is achieved at the expense of assuming that certain interactions between factors are negligible.

Figure 7.13 is an example of a fractional factorial design with two half-replicates. Either of these half-replicates could be used as fractional replicates requiring only eight experiments. The full design is a 32 factorial that would have required sixteen experiments.

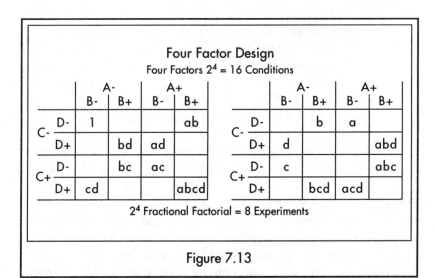

Four Factor Design
Four Factors $2^4 = 16$ Conditions

		A-		A+				A-		A+	
		B-	B+	B-	B+			B-	B+	B-	B+
C-	D-	1			ab	C-	D-		b	a	
	D+		bd	ad			D+	d			abd
C+	D-		bc	ac		C+	D-	c			abc
	D+	cd			abcd		D+		bcd	acd	

2^4 Fractional Factorial = 8 Experiments

Figure 7.13

Five Factor Design
Five Factors 2^5 = 32 Conditions

		A-				A+			
		B-		B+		B-		B+	
E	D	C-	C+	C-	C+	C-	C+	C-	C+
E-	D-	1			bc		ac	ab	
	D+		cd	bd		ad			abcd
E+	D-			bc	ac		ae		
	D+	de			bcde		acde	abde	

A One-Half Replicate of a 2^5 Factorial = 16 Experiments

Figure 7.14

Figure 7.14 is a plan for a five-factor design. If all combinations were run, there would be thirty-two experiments. In this example of a one-half replicate, only sixteen experiments were necessary.

For six factors at two levels (16 with half replication of six factors), the results are the following:
- n = 32;
- 3 plans of 2 blocks of 16 each;
- 4 blocks of 8 each; and
- 8 blocks of 4 each.

For seven factors at two levels (27 with one-fourth replication), these are the results:
- n = 32;
- 3 plans of with blocks of 16 each;
- 4 blocks of 8 each; and
- 8 blocks of 4 each.

Or,
- n = 16;
- 2 plans with 2 blocks of 8 each; and
- 4 blocks of 4 each.

These types of experiments should be replicated if all possible. See Appendix C for additional fractional factorial experimental designs.

2^n factorial experiments

In a 2^n factorial experiment, the n factors are studied, each of them on two levels. 2^n factorials are sensitive to bad values, which affect all other data in the experiment. Thus, factorial experiments should be replicated if at all possible. A conventional notation system is used to identify each of the experiments. A factor is identified by a capital letter, and its two levels by the subscripts zero and one. For example, if there are three factors, A, B, and C, the corresponding levels of the factors are A_0, A_1, B_0, B_1, and C_0, C_1, respectively. Customarily, the zero subscript refers to the lower level, the normal condition or the absence of condition, whichever is appropriate.

Figure 7.15 is an example of a 2^3 factorial experiment that uses blocks of four with two replicates. It is represented by a combination of small letters denoting the factor levels. These correspond with their uppercase counterparts when they are at the level denoted by the subscript one. Thus, in figure 7.16, the symbol a represents a treatment combination where A is at the level A_1, B is at B_0, and C is at C_0. In figure 7.15, the levels are designated as follows:

- for factor A:
  A_0 = Low Level,
  A_1 = High Level;
- for factor B:
  B_0 = Low Level,
  B_1 = High Level; and
- for factor C:
  C_0 = Low Level,
  C_1 = High Level.

2^n Factorial Experiment
2^3 Factor = 8 Experiments

	Level							
Factor A	Low	High	Low	High	Low	High	Low	High
Factor B	Low	Low	High	High	Low	Low	High	High
Factor C	Low	Low	Low	Low	High	High	High	High
Code	(1)	a	b	ab	c	ac	bc	abc

		A_0		A_1	
		B_0	B_1	B_0	B_1
C_0		(1)	b	a	ab
		N=2	N=2	N=2	N=8
C_1		c	bc	ac	abc
		N=2	N=2	N=2	N=8
		N=8	N=8	N=8	N=16

Figure 7.15

Factorial Experiment with Partial Confounding

Replicate 1		Replicate 2	
Block 1	Block 2	Block 1	Block 2
(1)	a	(1)	b
ab	b	a	c
ac	c	bc	ab
bc	abc	abc	ac

Figure 7.16

This type of coding defines blocks and identifies confounding (fig. 7.16). Confounding is the inability to distinguish between either a factor's main differential effects and the effects of other factor(s), block factor(s).

The estimate of BC can be obtained only from replicate 1 and that of ABC only from replicate 2. The remaining estimates for A, B, C, AB, and AC are obtainable using both replicates therefore will have greater precision.

Analysis of variance

Once an experiment has been designed, there are various methods of interpreting it. *Analysis of variance* has been found to be the optimal form of analysis in these situations. It subdivides the total variation of the data set into meaningful components associated with specific sources:

$$s \text{ Total} = \sqrt{s^2 \text{ } factors + s^2 \text{ } interactions + s^2 \text{ } residual}$$

Its purpose is to determine the contribution each factor makes to the variance of the table.

There are a number of methods for presenting the analysis of variance, the most common of which is the analysis of variance (Anova) table. Others are the sum of squares method and probability plot analysis. This section discusses the anova table and probability plot in detail. A simple factorial experiment is then introduced as an example of how to use the anova table and probability plots to interpret data.

Analysis of variance table

Analysis of variance (anova) tables are used to determine which variables are effective and which are not. The F-ratio in the table is also important because it shows the degree of relationships between the source variables and the dependent variables. An analysis of variance table is divided into columns for:

- source of variation: the factors being investigated in the experiment;
- the sum of squares (S.S.):

$$\Sigma(X - \overline{X})^2 = \Sigma X^2 - N\overline{X}^2;$$

- degrees of freedom (df): these are the same as the denominator n - 1 in the definition of sample variance;
- mean square (m.s.): the sum of squares divided by the degrees of freedom;
- F-ratio test of variances: the ratio of the experimental factor variance to the residual (or error) variance, or the factor mean square divided by the residual mean square; and
- significance: a test of the significance that decides whether the factor(s) being investigated differs from the residual error.

The application of statistical analysis involves an extensive amount of calculations, such as totaling each source of variation and its square. These calculations are a measure of variability (dispersion) of each observation, y, from its mean, y divided by the d.f., or

$$s^2 = \frac{\Sigma(y - y)^2}{n - 1}$$

The symbol y is used because it is considered a response variable. A computer uses a slightly different but equivalent formula:

$$s^2 = \frac{\left[\Sigma(y)^2 - \dfrac{\Sigma(y)^2}{n}\right]}{n - 1}, \text{ where}$$

s^2 = observation means square deviation,

y = data observation,

n = number of observations in the group,

n - 1 = degrees of freedom, and

Σ = summation symbol.

Designed Experiment Defect Data

Machines	I			II			III					
Materials	A	B	C	A	B	C	A	B	C	n	Σy	Σy²
Operator #1	21	19	10	19	16	12	21	10	11	9	139	2325
Operator #2	26	20	13	27	23	19	24	21	16	9	189	4137
Operator #3	19	16	15	18	14	11	23	15	13	9	144	2406
Σy	66	55	38	64	53	42	68	46	40	27	472	
Σy²	1478	1017	494	1414	981	626	1546	766	546			8868

Table 7.7

The following example illustrates the basic principles of anova tables. It is based on a 3 x 3 x 3 factorial experiment that also demonstrates the effects of all possible interactions.

A textile manufacturing plant was producing too many defective goods. Was the source of the defects the textile machinery, the operators, or the material purchased from different vendors? To determine an answer, three machines were selected to be run by three different operators with textile material from three major suppliers. Each combination was run for a full day, and the defects from each combination were tallied at the end of the day.

Table 7.7 shows the data collected for the twenty-seven different conditions. The analysis of this data was compiled in an analysis of variance table (table 7.8). The problem being studied, the major hypotheses, the outcomes for main variables and their interaction, the number of measurements taken, and the significance of the outcomes all may be discerned from this table.

The computations for the table involved separating the deviations of the observations from the arithmetic average into components relating to the different sources of variation.

First, the totals were grouped by machines and materials (columns) and operators (rows). All totals for the sets were checked to verify a grand total of 8868. The 26 degrees of freedom, one less than the 27 observations, were organized into 2 degrees of freedom each for differences between the 3 machines, the 3 materials, and the 3 operators. The interactions each had 4 degrees of freedom for differences between those factors that were paired:

- Machines-Materials (3-1)(3-1) = 4 d.f.;
- Machines-Operators (3-1)(3-1) = 4 d.f.; and
- Materials-Operators (3-1)(3-1) = 4 d.f.

Table 7.8 was then constructed by distinguishing the:

- adjustment for the mean needed to form the sums of squares;
- machine sum of squares;
- material sum of squares;
- operator sum of squares;
- machine-material interaction sum of squares;
- machine-operator interaction sum of squares;
- material operator interaction sum of squares;
- residual error sum of squares; and
- total sum of squares.

Analysis of Variance of Defect Data

(I) Source of Variation	(II) Sum of Squares	(III) Degrees of Freedom	(IV) Mean Square	(V) F Ratio	(VI) Signif.
Main Factors	y^2-ny^2	$n-1$	$\dfrac{y^2-ny^2}{n-1}$	s^2/s^2	%
Machines	1.8	2	.9	.2	none
Vendor Materials	339.8	2	169.9	36.1	.001
Operators	168.5	2	84.2	17.9	.01
Interactions					
Machines-Materials	18.4	4	4.6	1.0	none
Machine-Operators	40.4	4	10.1	2.1	none
Materials-Operators	10.3	4	2.6	.5	none
Residual Error	37.5	8	4.7		
Total	616.7	26			

Table 7.8

The adjustment for the mean needed to form the sum of squares is calculated with a computer:

$$s^2 = \frac{\left[\Sigma(y)^2 - \dfrac{\Sigma(y)^2}{n} \right]}{n-1}$$

This is also the mean sum of squares when there is no variance in any of the factors. It is called the correction factor and is subtracted from the other experimental factors.

The sum of squares calculations for the table produce the following results:

1. machine sum of squares, or the sum of defects for each machine squared and totaled:

$$\frac{\Sigma y^2}{n} = \frac{(159)^2 + (159)^2 + (154)^2}{9} = \frac{74289}{9}$$

$$= 8253.1 - 8251.3 = 1.8$$

2. material sum of squares, or the sum of defects for each material squared and totaled:

$$\frac{\Sigma y^2}{n} = \frac{(198)^2 + (154)^2 + (120)^2}{9} = \frac{77320}{9}$$

$$= 82591.1 - 8251.3 = 339.8$$

3. operator sum of squares, or the sum of defects for each operator squared and totaled:

$$\frac{\Sigma y^2}{n} = \frac{(198)^2 + (189)^2 + (144)^2}{9} = \frac{75778}{9}$$

$$= 8419.7 - 8251.3 = 165.5$$

4. machine-material interactions, or the sum of squares for each combined:

$$\Sigma y^2 = (66)^2 + (55)^2 + (38)^2 + (64)^2 + \dots (40)^2 = 25,384$$

$$\frac{\Sigma y^2}{n} = \frac{25834}{3} = 8611.3 - 8251.3 - 1.8 - 339.8 = 18.4$$

5. machine-operator interactions, or the sum of defects for each combined:

$$\Sigma y^2 = (50)^2 + (59)^2 + (50)^2 + (47)^2 + \ldots (51)^2 = 25,386$$

$$\frac{\Sigma y^2}{n} = \frac{25386}{3} = 8462.0 - 8251.3 - 1.8 - 168.5 = 40.4$$

6. material-operator interactions, or the sum of defects for each combined:

$$\Sigma y^2 = (61)^2 + (45)^2 + (33)^2 + (77)^2 + (64)^2 + (48)^2 + (60)^2 + (39)^2$$
$$= 26,310$$

7. the sum of squares of all deviations, across all the rows in table 7.9:

$$\Sigma y^2 = (21)^2 + (19)^2 + (10)^2 + \ldots + (23)^2 + (15)^2 + (13)^2 = 8868$$

Based on these calculations, the table tells the following story. Three main variables (factors) have three levels each. A portion of the total variance is associated with these main effects. There also are three first-order interactions. An interaction is a measure of how much the effect of one variable is determined by the value of another. For example, machines-operators accounts for a portion of the total variance. A first-order interaction occurs when two main effects are being studied. A second order interaction deals with three main effects. In this example, there is only one second order interaction: Machines-Materials-Operators. It is equal to the residual error because all the source variables are accounted for; it also accounts a portion of the total variance.

Which variables and/or interactions, then, are statistically significant? By dividing the source-mean square by the residual mean square for each variable, the F-value for each is obtained. With the degree of freedom for each variable as a guide, the F-table (table A.5 in Appendix A) may be used to find each effect's level of significance. Only two have a notable level of significance, vendor materials (.001) and operators (.01). The interactions all have no level of significance.

These results indicate that the machines were not the cause of the defects, nor were there interactions between the machines and operators. There was a big problem with the vendor materials and the operators, however. Material A should be eliminated and control charts should be placed at each operator's position to determine the reason for the significant differences.

Probability plot analysis of a 3 factor experiment

The previous section demonstrated how an analysis of variance table determines what variables are important and even the degree of their importance. However, an analysis of variance table cannot show the shape of the distributions and their relationships with each other. So, the data used to create table 7.8 will be examined in a probability plot, which might be considered the second part of an analysis of variance.

The data was ranked from the highest to the lowest values and recorded in a data collection sheet (table 7.9). From table A.1 in appendix A, the plotting positions for 27 and 9 data values were taken.

Data Collection Sheet

PART NUMBER	Product 24B	PART NAME	Material A, B & C	OPERATION	Sewing Blouses		SPECIFICATION LIMITS	0 Defects²	
OPERATOR #1, #2, & #3		MACHINE	Textile I, II & III	GAGE	Visual	UNIT OF MEASURE	Defects/Day	ZERO EQUALS	No Defects

Plotting Position i = 1	Table I Appendix 1.9%	Ranked Data 27		Plotting Position 1%	Machines			Materials			Operators		
					I°	II˟	III⁰	A°	B⁰	C˟	1˟	2⁰	3°
2	5.6	26		1 - 5.6	26	27	24	27	23	19	21	27	23
3	9.3	24		2-16.7	21	23	23	26	21	16	21	26	19
4	13.0	23		3-27.8	20	19	21	24	20	15	19	24	18
5	16.7	23		4-38.9	19	19	21	23	19	13	19	23	16
6	20.4	21		5-50.0	19	18	16	21	16	13	16	21	15
7	24.1	21		6-61.1	16	16	15	21	16	12	12	20	15
8	27.8	21		7-72.2	15	14	13	19	15	11	11	19	14
9	31.5	20		8-83.3	13	12	11	19	14	11	10	16	13
10	35.2	19		9-94.4	10	11	10	18	10	10	10	13	11
11	38.9	19											
12	42.6	19		∑ X	159	159	154	198	154	120	139	189	144
13	46.3	19		∑ X²	2989	3021	2858	4438	2764	1666	2325	4137	2406
14	50.0	18		X	17.7	17.7	17.1	22.0	17.1	13.3	15.4	21.0	16.0
15	53.7	16		σ	4.5	4.9	5.0	3.0	3.8	2.7	4.4	4.3	3.4
16	57.4	16											
17	61.1	16											
18	64.8	15											
19	68.5	15											
20	72.2	14											
21	75.4	13											
22	79.6	13											
23	83.3	12											
24	87.0	11											
25	90.7	11											
26	94.4	10											
27	98.1	10								PLANT 9	DEPT. 18	DATE	MBB PERFORMANCE SUPERVISOR	

Table 7.9

All data values were plotted on normal probability paper (fig. 7.17) and a line of best fit was drawn through them. When this line of best fit was compared with the data from table 7.11, this was the result:

$$y = \frac{472}{27} = 17.5$$

$$s^2 = \frac{616.7}{26} = 23.7$$

$$s = 4.86$$

$$\text{upper sigma} = \frac{17.5}{4.86} = 22.36$$

$$\text{lower sigma} = 17.5 - 4.86 = 12.64 \cdot$$

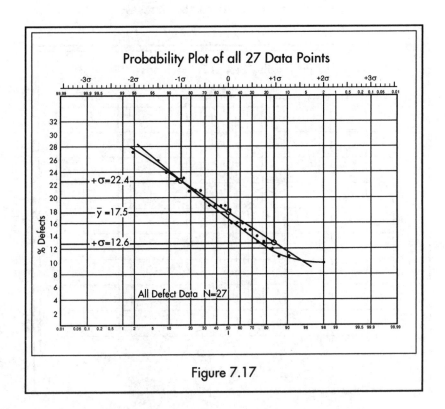

Figure 7.17

When a line was drawn through the y, the upper sigma, and the lower sigma value was compared with the line of best fit, the plotted data appeared to be bimodal. This was important information because the lower half of the probability curve indicated fewer defects.

A breakdown of probability plots by machine, material, and operators was then examined. Since there were nine data points (taken from table A.1 in appendix A), nine plotting points were necessary. The data points for each variable were ranked as shown in table 7.9. The results when each variable is compared with its position are shown in figure 7.18.

Like the analysis of variance table, the probability plot for the machines (fig. 7.18) showed no difference between the machines, nor was a possible machine-operator interaction shown to be statistically significant. The table also indicated that there was a significant difference between operators, but not *which* operators. During the charting of data for operators 1 and 3 on the probability plot, bimodal distributions for that data became evident. Therefore, the source of the significant differences was found to be operators 1 and 3.

This information was taken to a quality circle group that included machine operators and personnel from the purchasing department.

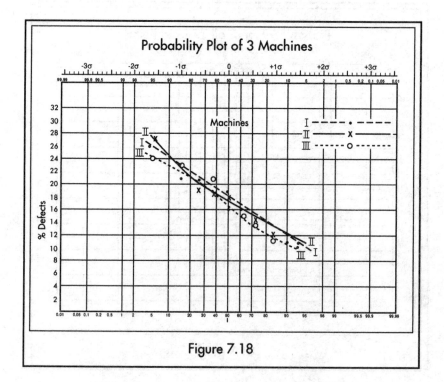

Figure 7.18

The data from table 7.8, when presented with a picture of the differences, supported the decision to eliminate vendors A and B and select vendor C. Operators 1 and 3 confirmed that they were aware of the differences in the materials and had made handling adjustments accordingly. The purchasing department agreed to purchase only material C and decided to ask the merchandising department to examine the specification requirements.

A factorial experiment from design to analysis

The following example demonstrates how a simple factorial experiment may be designed. It also examines several methods of conducting its analysis of variance.

A new induction heat treating machine was received from a manufacturer. The purchase order had specified that the machine was to meet OSHA standards. The machine, however, was so noisy that it failed to meet the OSHA specification of 90 decibels. The initial test measured 94 decibels, with a standard deviation of 2.0.

A simple factorial experiment was designed to discover the source of the noise and to determine if insulating part of the machine would reduce the noise to an acceptable level. The experiment is illustrated in figure 7.19.

Noise-Level Experiment

		Insulation Coils	Transformer	n	ΣX	ΣX^2
Location of Readings	End	$n = 10$ $\bar{X} = 91.0$ $\Sigma X = 910$ $\Sigma X^2 = 82840$	$n = 10$ $\bar{X} = 88.4$ $\Sigma X = 884$ $\Sigma X^2 = 78166$	20 1794 161006		
		I	II			
	Front	$n = 10$ $\bar{X} = 90.1$ $\Sigma X = 901$ $\Sigma X^2 = 81193$	$n = 10$ $\bar{X} = 87.6$ $\Sigma X = 876$ $\Sigma X^2 = 76752$	20 1777 157945		
		III	IV			
	n	20	20	40		
	ΣX	1811	1760	3571		
	ΣX^2	164033	154918	318951		

Figure 7.19

Noise Level Data						
		n	$\bar{X}$	ΣX	ΣX^2	s
I	88,92,92,90,93,91,94,91,89,90	10	91.0	910	82840	1.8
II	89,91,89,88,87,89,90,88,86,87	10	88.4	884	78166	1.5
III	89,92,91,90,91,91,90,90,89,88	10	90.1	901	81193	1.2
IV	89,90,87,88,88,86,86,87,87,88	10	87.6	876	76752	1.3

Table 7.10

Using the calculations described and the noise-level statistical data from table 7.10, the experimenters prepared an analysis of variance table (table 7.11). An examination of table 7.11 revealed that insulating the machine was indeed a significant factor in reducing the noise level. The source of the noise, however, remained problematic.

The experimenters decided that the data distribution should be scrutinized in a probability plot, especially since it approached the OSHA specifications. Statistical data from table 7.11 was ranked for each factor from the highest to the lowest value (table 7.12), and the procedures outlined in chapter 5 for constructing a probability plot were followed. The results are illustrated in figure 7.20.

Analysis of Variance Table					
(I) Source of Variation	(II) Sum of Squares	(III) Degrees of Freedom	(IV) Mean Square	(V) F Ratio	(VI) Signif.
	$\Sigma X^2 - n\bar{X}^2$	n-1	$\dfrac{\Sigma X^2 - n\bar{X}^2}{n-1}$	$\sigma^2/\sigma^2_{resid.}$	%
Main Factors					
Insulation-*Columns*	65.0	1	65.0	30.9	.001
Location-*Rows*	7.2	1	7.2	3.4	.01
1st Order Interaction	—	—	—	—	n.s.
Residual	77.8	37	2.1		
Total Variance	150.0	39	5.5		

Table 7.11

Probability Plotting Data for Noise Level
Ranked Data on Four Factors of Noise Level

Rank	Plotting Point %	I	II	III	IV
1	5.0	94	91	92	90
2	15.0	93	90	91	89
3	25.0	92	89	91	88
4	35.0	92	89	91	88
5	45.0	91	89	90	88
6	55.0	90	88	90	87
7	65.0	90	88	90	87
8	75.0	90	87	89	87
9	85.0	89	87	89	86
10	95.0	88	86	88	86
$\bar{X}$	50.0	91.0	88.4	90.1	87.6
Plotting Symbols		x	+	o	#

Table 7.12

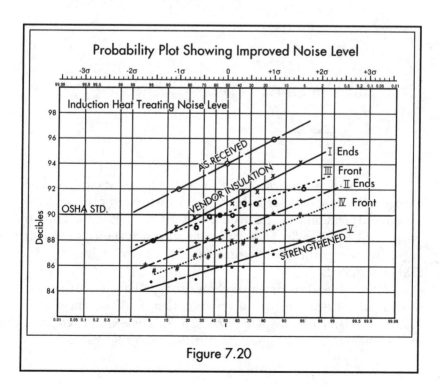

Figure 7.20

The probability plots demonstrated that insulating the coils (I and III in the illustration) significantly improved the noise level but not enough to meet OSHA standards. Insulating the transformer (II and IV in the illustration) also lowered the noise level, but about 10 percent of the noise was still above OSHA standards.

Experimenters noted that the readings from the front of the machine (III and IV in the illustration) had less variability than the ends (I and II). They assumed that the front of the machine had a firmer mechanical structure than the ends holding the coils. They decided that a structural support of the ends would reduce the noise level even more.

The end of the beam holding the coils and the beam itself were strengthened. With the insulation of the coils and transformer still in place, and additional ten readings were taken with the following results:

$$V = 86, 88, 85, 86, 85, 86, 87, 86, 87, 85$$

$$n = 10$$

$$\overline{X} = 86.1$$

$$\Sigma X = 861$$

$$\Sigma X^2 = 74141$$

$$s = .99$$

The machine was then able to meet OSHA standards. The change in the slope of the probability plot (V), when compared with the as-received condition and the vendor's insulation plots (I and II), was notable. The probability plots then were successfully used in vendor negotiations regarding reimbursement of the costs for meeting the OSHA standards.

Although the experimenters chose to use probability plots for the second part of their analysis of variance, they could also have used multiple box-and-whisker plots (see Chap. 5). These are also effective in the analysis of a designed experiment, especially when factors

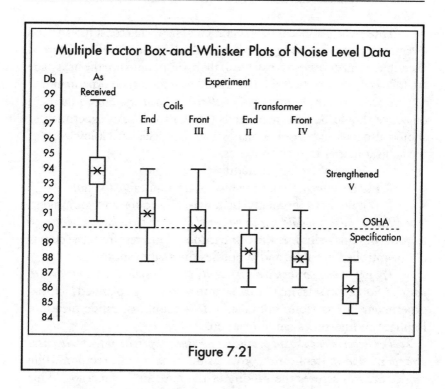

Figure 7.21

characterized by a limited amount of data are being compared. They can also compare more factors than a probability plot.

In figure 7.21, the data for the noise-reduction experiment is plotted as a multiple-box-and-whisker plot. The factors shown to be significant in the analysis of variance table are graphically evident.

Box-and-whisker plots can be problematic in their imprecision, however. The plots are based on quarters and their results are not very sensitive. The same can be said of their relationship to specifications, as the percentage of the product out of specification cannot be determined. This may be why, unless there are many factors, probability plots are the preferred graphical method for analysis of variance.

Conclusion

This chapter has introduced the use of designed experiments and their analysis as viable methods for process study and improvement. Several examples of designed experiments and designed factorial experiments have been given, and the latter have been shown to be particularly useful in industrial situations.

Those who planned these experiments knew what constitutes a good designed experiment. However, those embarking on their first designed experiments may not be so fortunate. If the basics outlined on the next page are followed, a researcher will have a solid base for a good experiment.

A good experiment must have a clearly defined objective. The main objective should be customer needs (see Chap. 3). An experimenter should also have at his or her disposal all the specialized knowledge of the subject matter in such things as:

- choice of factors, including their range;
- knowledge of what the results are applicable to; and
- choice of experimental materials, procedure, and equipment.

The effects of factors should not be obscured by other variables. The use of factorial designs helps free the comparisons of interest from the effects of uncontrolled variables and simplifies the analysis of the results.

The experimenter should be as free of bias (conscious or unconscious) as possible. Some variables may be taken into account by a planned factorial experiment For variables not taken into account, use randomization. Replication fine-tunes randomization.

The experiment should provide a measure of precision (experimental error or residuals). Replication provides this measure of precision while randomization assures the validity of the measure of precision. This precision must also meet the experiment objectives.

Greater precision may be achieved by refined techniques, such as control charts, probability plots and replications. In industrial experimentation, planned factorial groupings may be used to take advantage of naturally homogeneous groupings in materials, machines, shifts, time, and even other production lines or plants that are not direct factors in the experiment.

Increasing the sample size also increases the precision of the sample data. However, an increase in precision is not in direct proportion to the sample size, but only to the square root of the sample. Instead of increasing the sample size, process changes should be introduced, and samples taken of them.

Once a good designed experiment is planned, implemented, and analyzed, a process capability study is possible. Chapter 8 discusses how to evaluate a process as the final step in the plan for investigation.

Chapter 8

Evaluation

"In controlling the quality of manufactured product, it is one thing to measure the quality to see whether or not it meets certain standards; and it is quite another thing to make use of these measurements to predict and control the quality in the future."

— Walter Shewhart

Evaluating a manufacturing process involves not only data gathering and data analysis, but also process specifications. The statistic used in the evaluation process is the Cpk index, which includes both specification and process variation data.

A probability plot should be created first for an overall view of the process in relationship to the specifications. A process control chart is the best tool for the process capability studies, with the R-chart being the most important chart in the study.

The R-chart represents the process's inherent variation, which is predetermined by the manufacturing process; these are the items that are basically controlled by management, including machines, tools, materials, process, maintenance, fixtures, and operating training.

Specifications ensure a product's function to meet the needs and expectations of the customer. The specifications are the requirements for materials, finishes, tolerances, functions, and so forth. These factors are usually expressed on a drawing as target dimensions with a high and low limit or tolerances. Quality control ensures that the part is made not only to conform but also to surpass these specifications.

When meeting specifications becomes difficult, an effort should be made to find if the specifications are realistic. If not, there should be an investigation to determine if the specifications could be altered. Loosening specifications without the process being in a state of statistical control should be done with caution, however. Otherwise, the product or process may deteriorate.

On the other hand, there are functional and overall cost advantages to manufacturing a product based on tighter specifications. The secret of zero defects accounted for the success of the Japanese in industry.

Requirements for industrial specifications

There are certain aspects of an industrial specification that should always be verified. To determine them, the following questions may be asked:

- are the specifications reasonable? Unnecessary exactitude is expensive, as is unnecessary detail;
- can they be verified?
- can several vendors meet them? Competition must always be considered;
- are they clear and understandable; and
- are they flexible within the limits of the functional needs?

Specifications should be rigid enough to ensure quality, but flexible enough to allow improvement and progress.

Identifying limits

When evaluating specifications, the differences between specification limits, process limits, and control chart limits should be distinguished (fig. 8.1). Each has a different meaning; they are not interchangeable. For example, it is possible to have a sample of five points that are out of the control limits but within process limits and within specification limits.

Specification limits are specified as a process requirement.

Control chart limits ($\pm A_2 R$) are based on samples from a process capability study and are used to determine whether a process is in or out of control. Process limits are the natural six sigma limits of the operating process and are usually associated with gauge limits. Process limits are wider than control chart limits and are less then, equal to, or even possibly greater than specification limits (see fig. 8.4).

Control charts and probability plots demonstrate how well centered the process is in respect to specifications. When specification limits are

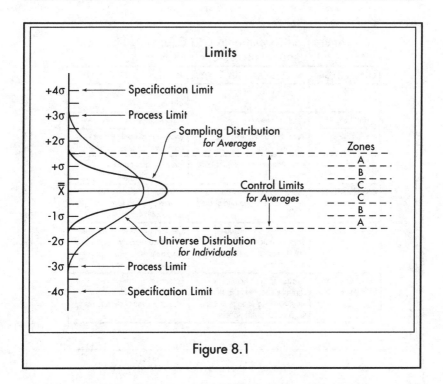

Figure 8.1

drawn on probability plots, there is an added advantage. The percentage of the product or process that is out of specification is directly evident. Since the probability plot shows the shape of the distribution, the effects of the distribution on the specification are also evident.

It is possible, but not very probable, to have a sample of five data points that are out of control limits but within process limits and within specification limits (see fig. 8.4).

Process capability study

Process capability is the inherent ability of a manufacturing process to produce similar items that can be controlled for a sustained period of time, given a certain set of conditions. It is compared to a set of given specification limits.

The purpose of a *process capability study* is to determine the limits within which a process operates. This is based on minimum variability, which is governed by either prevailing circumstances or the process's inherent variability. The study is systematic and uses probability plots and process control charts.

Probability plots are used to obtain information about the process distribution and how the distribution relates to the specifications.

	Difference Between a Process Capability Study and a Process Control Chart	
CHARACTERISTIC	PROCESS CAPABILITY STUDY	PROCESS CONTROL CHART
PURPOSE	To obtain information	To maintain a predetermined distribution
SAMPLES	Relatively few	A running series
ANALYSIS	Very careful analysis and interpretation	Shop watches only the more obvious changes in pattern
ACTION	Any change may be important, either good or bad	The shop acts only on unwanted changes
INFORMATION	Distribution shape is studied as well as average and spread	Attention focused mainly on average and spread (or percent defectives)
CENTERLINES	Centerlines are calculated from the data to reflect the distribution of the process being studied	Centerlines are set to represent a balance between quality and cost; they show where we want the process to run
RELATION TO SPECIFICATION	Relation to specification is carefully checked. The study may lead to a change in either the process or the specification	Proper relationship to specification is allowed for when the control chart is set up

Figure 8.2

Process control charts are used to discern whether the process is in statistical control and behaving naturally or unnaturally; they also help determine the cause(s) of unnatural behavior, in real time. Control charts are more of a manifestation of process control, however, and cannot replace a process capability study. The differences are illustrated in figure 8.2.

The natural behavior of the process after the unnatural behavior is eliminated is another definition of process capability, and its measure is the *Cpk index*. To make improvements in the natural variation, a designed experiment (see Chap. 7) and an analysis of variance are used, for they determine the source(s) of these variations.

When the causes of the process variation have been identified, one of the following courses of action should be taken on the processes:

- maintain (this is continuing inspection to remove defectives);
- eliminate (this improves the process);
- reduce (to an economical level); and
- advantageous, deliberate use (to control the process).

Experimenters should always try to reduce the specifications. Opening them up will cause variability to increase.

Process capability procedure

To carry out a process capability study, these steps should be taken:

1. Identify the process, such as machining, assembly, or product.

2. Identify the key operations, such as milling or broaching.

3. Identify the controlling parameters, such as speeds, temperature or time.

4. Determine the sample size. The subgroup size is usually five. The subgroup samples are taken from the process on a random schedule, from five minutes up to several hours. Try to get at least twenty groups of subgroups. Depending on the process, it may be necessary to take only one subgroup per day. Carefully record the times the samples are taken.

5. Measure the key operations or dimensions.

6. Analyze the measured data with: probability plots, control charts, and designed experiments.

7. Determine if the process is in statistical control.

8. Determine the estimated process standard deviation.

9. Determine the Cpk index for both the upper and lower specifications.

Statistical process improvement is a continuing process. The probability plot will indicate if the data is normally distributed. The designed experiment will help determine the significant factors. If the process is not in statistical control, trying to determine the assignable causes of the variation is the next logical step. The process control chart will indicate if there are any assignable causes. If the process requires a tighter distribution after the assignable causes are determined (Cpk less than 1.00), then a designed experiment and an analysis of variance are necessary. If corrective action is taken, experimenters should return to the fourth step, determine a sample size, and go through the process again to verify new results.

Many of the procedures listed above have been discussed in previous chapters. Several, however, need to be defined or discussed in greater detail.

Estimated process standard deviation

There are three ways of obtaining the estimated process standard deviation:

1. First, it may be calculated by a calculator or a computer from the standard deviation equation:

$$\hat{\sigma} = \sqrt{\frac{(\Sigma X^2) - \frac{(\Sigma X)^2}{n}}{n-1}}, \text{ where}$$

   x = data values, and

   n = the number of data values in the study.

2. It may be taken from the R-process control chart. The R from the chart, divided by the d_2 factor for the sample size (n) used to compute the R-bar (usually n = 5):

$$\hat{\sigma} = \frac{\overline{R}}{d_2}$$

   For 5, d_2 = 2.326.

3. It may also come from a probability plot as the difference between the mean (X) and the 16 percent and/or the 84 percent value (see fig. 5.19).

Process capability index

The *capability index*, a measure of the process capability, is a ratio of the specifications to the process variation. Mathematically, this is written as:

$$CP = \frac{Specification\ Limits}{Process\ Variation} = \frac{USL - LSL}{6\sigma}$$

   where: USL = upper specification limit,

   LSL = lower specification limit, and

   6σ = estimated process variation.

Determining the ratio of the specifications to the process variation in turn determines the acceptable process variability. Since the natural variability of the process can be described by the normal distribution, capability can be evaluated by comparing distribution properties to the process specifications with the index Cpk.

The *Cpk index* determines the capability in relation to the *specification mean* and involves both the upper and lower specification limits separately. It is based on the "worst-case" view of the data.

To determine the Cpk values, the process mean ($\overline{\overline{X}}$) and the standard deviation (σ) of the process must be obtained. The lesser value of the difference between the process mean and the upper or lower specification limit is used.

$$Cpk = \frac{Upper\ Specification\ Limit - Process\ Mean}{3\sigma} = \frac{USL - \overline{X}}{3\sigma}$$

$$Cpk = \frac{Process\ Mean - Lower\ Specification\ Limit}{3\sigma} = \frac{\overline{X} - LSL}{3\sigma}$$

Evaluating the process capability study

Once the Cpk values have been obtained, the status of the process may be determined. Processes that are in control and capable have been shown to have a Cpk of 1.33 or greater. This is comparable to having eight sigma specification limits (fig. 8.3).

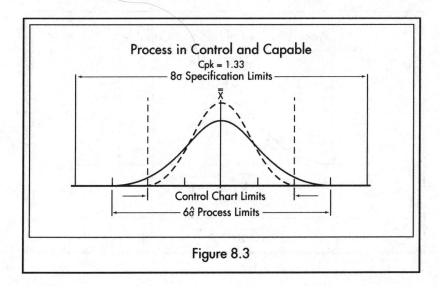

Figure 8.3

A Cpk between 1.00 and 1.33 means a process may be making good parts and still not be in control (fig. 8.4). It should be monitored more closely.

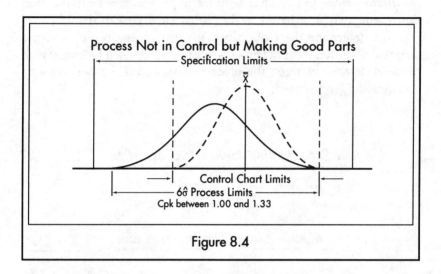

Figure 8.4

A process that is in control and but not capable (fig. 8.5) will have a Cpk less than 1.00. This means that the process limits are beyond the specification limits. A Cpk less than 1.00 also may indicate a process not in control and not capable (fig. 8.6). When a process is not in control and not capable, it is producing defects.

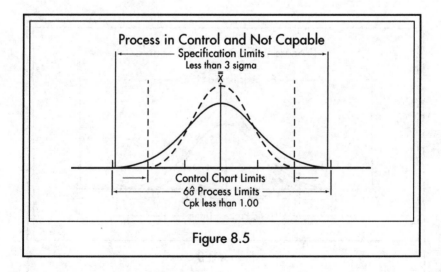

Figure 8.5

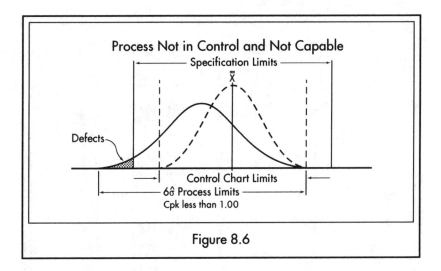

Figure 8.6

Cpk management report

The Cpk index is a useful tool for communications with management, since it is an overall measure of process capability. It requires specification limits and process standard deviation for its calculation. It also reduces the ratio of the specifications to the process variables to a single number, providing managers with an index of performance. The index should be reported on a regular basis, particularly as new characteristics are added, for the indexes will change and some will no longer be necessary after time.

An example of an SPC production report is shown in figure 8.7. The data was collected by part number, part name, operation number, and machine number. The values of most interest are the upper and lower Cpks, which appear in the last two columns of the report. Upon evaluation of the values as outlined in the previous section, a rough measure of the process capability may be determined.

Other data, which is useful for indicating the thoroughness of the study, appearing in this report are:
- characteristic number;
- number of data points calculated;
- data start and ending dates;
- number of points above and below 3 sigma limit;
- number of points above and below upper specification limit;
- drawing specifications; and
- upper and lower Cpk.

Statistical Process Control Management Report

DEPT. 3103

MANAGEMENT REPORT — STATISTICAL PROCESS CONTROL DATA

CHAR. NUMB.	NO PTS	DATE FROM	DATE TO	X-BAR	R-BAR	UCLX	LCLX	PTS ABV 3S	PTS BLW 3S	PTS ABV USL	PTS BLW LSL	SPECIFICATION	USL	LSL	UPPER CPK	LOWER CPK
** PART NUM: 12322017-SA PART NAME: RACE BRG., INNER OP NO: 60 MACH. NUM: 341501331																
* DATE 05/20/88																
081	15	0510	0518	-0.00105	0.00090	-0.00053	-0.00157	0	0	0	5	88.744 +.000 -.001	0.00000	-0.00100	0.88	-0.04
SPC1	15	0510	0518	-0.00140	0.00150	-0.00053	-0.00227	0	0	0	0	88.060 +.000	0.00000	-0.00250	0.70	0.55
** PART NUM: 12343976 PART NAME: LARGE OUTER RING OP NO: 295 MACH. NUM: 341501332																
* DATE 05/20/88																
010	20	0517	0517	0.00115	0.00475	0.00389	-0.00159	1	0	8	1	87.263 +.000	0.00000	-0.00075	-0.18	0.30
** PART NUM: 12343976 PART NAME: OUTER RING OP NO: 60 MACH. NUM: 341106723																
* DATE 05/13/88																
13	20	0411	0503	-0.00010	0.00237	0.00130	-0.00150	0	0	0	0	3.965 +- .005	0.00500	-0.00500	1.60	1.54
14	20	0411	0503	-0.00020	0.00155	0.00110	-0.00070	0	2	0	0	88.2175 +- .0013	0.00130	-0.00130	0.53	0.72
16	20	0411	0503	-0.00070	0.00300	0.00100	-0.00240	1	0	0	0	.345 +- .005	0.00500	-0.00500	1.42	1.07
17	20	0411	0503	0.00002	0.00287	0.00170	-0.00160	0	0	0	0	89.9 +- .015	0.01500	-0.01500	3.90	3.91
19	20	0411	0503	0.00107	0.00275	0.00270	-0.00050	0	0	1	0	87.31 +- .0025	0.00250	-0.00250	0.39	0.97
20	20	0411	0503	0.00005	0.00175	0.00110	-0.00100	0	0	0	0	93.72 +- .015	0.01500	-0.01500	6.39	6.43
21	20	0411	0503	-0.00115	0.00137	-0.00040	-0.00190	0	3	0	0	3.395 +- .005	0.00500	-0.00500	3.36	2.10
30	20	0411	0503	-0.00070	0.00087	-0.00020	-0.00120	0	0	0	0	2.8 +- .005	0.00500	-0.00500	4.87	3.68
** PART NUM: 12343978 PART NAME: RACE ASSY., INNER OP NO: 40 MACH. NUM: 341503337																
* DATE 05/20/88																
001	20	0510	0518	0.00000	0.00400	0.00231	-0.00231	1	1	2	2	89.7420 +.00075 -.0000	0.00075	0.00000	0.14	0.00
007	20	0510	0518	0.00005	0.00025	0.00019	-0.00009	1	0	0	0	2.125 +- .005	0.00500	-0.00500	15.00	15.30

Figure 8.7

Conclusion

A thorough process capability study incorporates all the procedures and components outlined in this chapter and demonstrated by the report in figure 8.7. The most important characteristic of the study is that it and its evaluation are ongoing processes. Production and manufacturing processes and their specifications may always be refined. If solutions to problems are suggested, they must be tested in the same manner as the processes before them. Such actions can only lead to process improvement, quality products, and customer satisfaction.

Appendix A

Important Tables

Plotting Positions for Probability Plots

Plotting Positions $100(i-0.5)/n$

Columns are indexed by n (values 6–20); rows by i.

i	6	7	8	9	10	11	12	13	14	15	16	17	18	19	20	i
1	8.3	7.1	4.2	5.6	5.8	4.5	4.2	3.8	3.6	3.3	3.1	2.9	2.8	2.6	2.5	1
2	25.8	21.4	18.7	16.7	15.8	13.6	12.5	11.5	18.7	18.8	9.4	8.8	8.3	7.9	7.5	2
3	41.7	35.7	31.2	27.8	25.8	22.7	28.8	19.2	17.9	16.7	15.6	14.7	13.9	13.2	12.5	3
4	58.3	58.8	43.7	38.9	35.8	31.8	29.2	26.9	25.8	23.3	21.9	28.6	19.4	18.4	17.5	4
5	75.8	64.3	56.2	58.8	45.8	48.9	37.5	34.6	32.1	38.8	28.1	26.5	25.8	23.7	22.5	5
6	91.7	78.6	68.7	61.1	55.8	58.8	45.8	42.3	39.3	36.7	34.4	32.4	38.6	28.9	27.5	6
7		92.9	81.2	72.2	65.8	59.1	54.2	58.8	46.4	43.3	48.4	38.2	36.1	34.2	32.5	7
8			93.7	83.3	75.8	68.2	62.5	57.7	53.6	58.8	46.9	44.1	41.7	39.5	37.5	8
9				94.4	85.8	77.3	78.8	65.4	68.7	56.7	53.1	98.8	47.2	44.7	42.5	9
10					95.8	86.4	79.2	73.1	67.9	63.3	59.4	55.9	52.8	50.0	47.5	10
11						95.5	87.5	88.8	75.8	78.8	65.6	61.8	58.3	55.3	52.5	11
12							95.8	88.5	82.1	76.7	71.9	67.6	63.9	60.5	57.5	12
13								96.2	89.3	83.3	78.1	73.5	69.4	65.8	42.9	13
14									96.4	90.0	84.4	79.4	75.8	71.1	67.5	14
15										96.7	98.4	85.3	88.6	76.3	72.5	15
16											96.9	91.2	86.1	81.6	77.5	16
17												97.1	91.7	86.8	82.5	17
18													97.2	92.1	87.5	18
19														97.4	92.5	19
20															97.5	20
i	6	7	8	9	10	11	12	13	14	15	16	17	18	19	20	i

Columns are indexed by n (values 21–35); rows by i.

i	21	22	23	24	25	26	27	28	29	30	31	32	33	34	35	i
1	2.4	2.3	2.2	2.1	2.8	1.9	1.9	1.8	1.7	1.7	1.6	1.6	1.5	1.5	1.4	1
2	7.1	6.8	6.5	6.2	6.8	5.8	5.6	5.4	5.2	5.8	4.8	4.7	4.5	4.4	4.3	2
3	11.9	11.4	10.9	10.4	10.6	9.6	9.3	8.9	8.6	8.3	8.1	7.8	7.6	7.4	7.1	3
4	16.7	15.9	15.2	14.6	14.0	13.5	13.0	12.5	12.1	11.7	11.3	10.9	10.6	10.3	10.0	4
5	21.4	20.5	19.6	18.7	18.0	17.3	16.7	16.1	15.5	15.0	14.5	14.1	13.6	13.2	12.9	5
6	26.2	25.8	23.9	22.9	22.0	21.2	20.4	19.6	19.0	18.3	17.7	17.2	16.7	16.2	15.7	6
7	31.0	29.5	28.3	27.1	26.0	25.0	24.1	23.2	22.4	21.7	21.0	20.3	19.7	19.1	18.6	7
8	35.7	34.1	32.6	31.2	30.0	28.8	27.8	26.8	25.9	25.0	24.2	23.4	22.7	22.7	21.4	8
9	40.5	38.6	37.8	35.4	34.0	32.7	31.5	30.4	29.3	20.3	27.4	26.6	25.8	25.0	24.3	9
10	45.2	43.2	41.3	39.6	38.0	36.5	35.2	33.9	32.8	31.7	30.6	29.7	28.8	27.9	27.1	10
11	50.0	47.7	45.7	43.7	42.8	40.4	38.9	37.5	36.2	35.0	33.9	32.8	31.8	30.9	30.0	11
12	54.8	52.3	58.8	47.9	46.8	44.2	42.6	41.1	39.7	38.3	37.1	35.9	34.8	13.8	32.9	12
13	59.5	56.8	54.3	52.1	50.0	48.1	46.3	44.6	43.1	41.7	40.3	39.1	37.9	36.8	35.7	13
14	64.3	61.4	58.7	56.2	54.0	51.9	50.0	49.2	46.6	45.8	43.5	42.2	40.9	39.7	38.6	14
15	69.8	65.9	63.0	60.4	58.0	55.8	53.7	51.8	50.0	48.3	46.8	45.3	43.9	42.6	41.4	15
16	73.8	70.5	67.4	64.6	62.0	59.6	57.4	55.4	53.4	51.7	50.0	48.4	47.0	45.6	44.3	16
17	78.6	75.0	71.7	68.7	66.0	63.5	61.1	58.9	56.9	55.0	53.2	51.6	50.0	48.5	47.1	17
18	83.3	79.5	76.1	72.9	70.0	67.3	64.8	62.5	60.3	58.3	56.5	54.7	53.8	51.5	50.0	18
19	88.1	84.1	80.4	77.1	74.0	71.2	68.5	66.1	63.8	61.7	59.7	57.8	56.1	54.4	52.9	19
20	92.9	88.6	84.8	81.2	78.8	75.8	72.2	69.6	67.2	65.8	62.9	69.9	59.1	57.4	55.7	20
21	97.6	93.2	89.1	85.4	82.0	78.0	75.9	73.2	70.7	68.3	66.1	64.1	62.1	60.3	58.6	21
22		97.7	93.5	89.6	86.0	82.7	79.6	76.8	74.1	71.7	69.4	67.2	65.2	63.2	61.4	22
23			97.8	93.7	98.8	86.5	83.3	88.4	77.6	75.8	72.6	78.3	68.2	66.2	64.3	23
24				97.9	94.0	90.4	87.0	83.9	81.8	78.3	75.8	73.4	71.2	69.1	67.1	24
25					98.8	94.2	90.7	87.5	84.5	81.7	79.8	76.6	74.2	72.1	70.0	25
26						98.1	94.4	91.1	87.9	85.8	82.3	79.7	77.3	75.0	72.9	26
27							98.1	94.6	91.4	88.3	85.5	82.8	80.3	77.9	75.7	27
28								98.2	94.8	91.7	88.7	85.9	83.3	80.9	78.6	28
29									98.3	95.8	91.9	89.1	84.4	83.8	81.4	29
30										98.3	95.2	92.2	89.4	86.8	84.3	30
31											98.4	95.3	92.4	89.7	87.1	31
32												98.4	95.5	92.4	98.8	32
33													98.5	95.6	92.9	33
34														98.5	95.7	34
35															98.6	35
i	21	22	23	24	25	26	27	28	29	30	31	32	33	34	35	i

From Volume 1: How to Analyze Data with Simple Plots
ASQC Basic References in Quality Control. Wayne Nelson

Table A.1

Standard Normal Distribution   Normal Distribution

Pz = the proportion of process output beyond a particular value of interest (such as a specification limit) that is z standard deviation units away from the process average (for a process that is in statistical control and is normally distributed). For example, if z = 2.17, Pz = .0150 or 1.5%. In any actual situation, this proportion is only approximate.

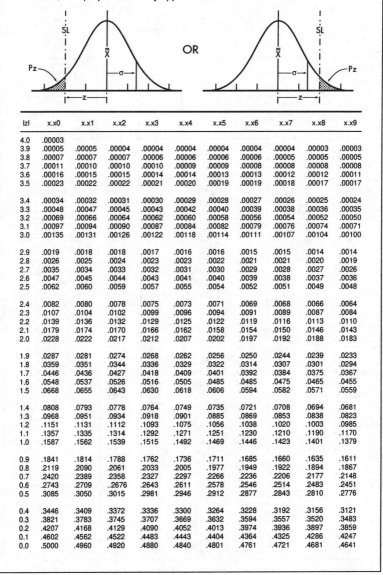

|z|	x.x0	x.x1	x.x2	x.x3	x.x4	x.x5	x.x6	x.x7	x.x8	x.x9
4.0	.00003									
3.9	.00005	.00005	.00004	.00004	.00004	.00004	.00004	.00004	.00003	.00003
3.8	.00007	.00007	.00007	.00006	.00006	.00006	.00006	.00005	.00005	.00005
3.7	.00011	.00010	.00010	.00010	.00009	.00009	.00008	.00008	.00008	.00008
3.6	.00016	.00015	.00015	.00014	.00014	.00013	.00013	.00012	.00012	.00011
3.5	.00023	.00022	.00022	.00021	.00020	.00019	.00019	.00018	.00017	.00017
3.4	.00034	.00032	.00031	.00030	.00029	.00028	.00027	.00026	.00025	.00024
3.3	.00048	.00047	.00045	.00043	.00042	.00040	.00039	.00038	.00036	.00035
3.2	.00069	.00066	.00064	.00062	.00060	.00058	.00056	.00054	.00052	.00050
3.1	.00097	.00094	.00090	.00087	.00084	.00082	.00079	.00076	.00074	.00071
3.0	.00135	.00131	.00126	.00122	.00118	.00114	.00111	.00107	.00104	.00100
2.9	.0019	.0018	.0018	.0017	.0016	.0016	.0015	.0015	.0014	.0014
2.8	.0026	.0025	.0024	.0023	.0023	.0022	.0021	.0021	.0020	.0019
2.7	.0035	.0034	.0033	.0032	.0031	.0030	.0029	.0028	.0027	.0026
2.6	.0047	.0045	.0044	.0043	.0041	.0040	.0039	.0038	.0037	.0036
2.5	.0062	.0060	.0059	.0057	.0055	.0054	.0052	.0051	.0049	.0048
2.4	.0082	.0080	.0078	.0075	.0073	.0071	.0069	.0068	.0066	.0064
2.3	.0107	.0104	.0102	.0099	.0096	.0094	.0091	.0089	.0087	.0084
2.2	.0139	.0136	.0132	.0129	.0125	.0122	.0119	.0116	.0113	.0110
2.1	.0179	.0174	.0170	.0166	.0162	.0158	.0154	.0150	.0146	.0143
2.0	.0228	.0222	.0217	.0212	.0207	.0202	.0197	.0192	.0188	.0183
1.9	.0287	.0281	.0274	.0268	.0262	.0256	.0250	.0244	.0239	.0233
1.8	.0359	.0351	.0344	.0336	.0329	.0322	.0314	.0307	.0301	.0294
1.7	.0446	.0436	.0427	.0418	.0409	.0401	.0392	.0384	.0375	.0367
1.6	.0548	.0537	.0526	.0516	.0505	.0485	.0485	.0475	.0465	.0455
1.5	.0668	.0655	.0643	.0630	.0618	.0606	.0594	.0582	.0571	.0559
1.4	.0808	.0793	.0778	.0764	.0749	.0735	.0721	.0708	.0694	.0681
1.3	.0968	.0951	.0934	.0918	.0901	.0885	.0869	.0853	.0838	.0823
1.2	.1151	.1131	.1112	.1093	.1075	.1056	.1038	.1020	.1003	.0985
1.1	.1357	.1335	.1314	.1292	.1271	.1251	.1230	.1210	.1190	.1170
1.0	.1587	.1562	.1539	.1515	.1492	.1469	.1446	.1423	.1401	.1379
0.9	.1841	.1814	.1788	.1762	.1736	.1711	.1685	.1660	.1635	.1611
0.8	.2119	.2090	.2061	.2033	.2005	.1977	.1949	.1922	.1894	.1867
0.7	.2420	.2389	.2358	.2327	.2297	.2266	.2236	.2206	.2177	.2148
0.6	.2743	.2709	..2676	.2643	.2611	.2578	.2546	.2514	.2483	.2451
0.5	.3085	.3050	.3015	.2981	.2946	.2912	.2877	.2843	.2810	.2776
0.4	.3446	.3409	.3372	.3336	.3300	.3264	.3228	.3192	.3156	.3121
0.3	.3821	.3783	.3745	.3707	.3669	.3632	.3594	.3557	.3520	.3483
0.2	.4207	.4168	.4129	.4090	.4052	.4013	.3974	.3936	.3897	.3859
0.1	.4602	.4562	.4522	.4483	.4443	.4404	.4364	.4325	.4286	.4247
0.0	.5000	.4960	.4920	.4880	.4840	.4801	.4761	.4721	.4681	.4641

Table A.2

t-Distribution of Critical Values

Central area —
T curve

Lower-tail area —
— Upper-tail area

-t critical value   0   t critical value

z critical values	1.28	1.645	1.96	2.33	2.58	3.09	3.29
Level of significance for a two-tailed test	.20	.10	.05	.02	.01	.002	.001
Level of significance for a one-tailed test	.10	.05	.025	.01	.005	.001	.0005

	1	3.08	6.31	12.71	31.82	63.66	318.31	636.62
	2	1.89	2.92	4.30	6.97	9.93	23.33	31.60
	3	1.64	2.35	3.18	4.54	5.84	10.21	12.92
	4	1.53	2.13	2.78	3.75	4.60	7.17	8.61
	5	1.48	2.02	2.57	3.37	4.03	5.89	6.86
	6	1.44	1.94	2.45	3.14	3.71	5.21	5.96
	7	1.42	1.90	2.37	3.00	3.50	4.79	5.41
	8	1.40	1.86	2.31	2.90	3.36	4.50	5.04
	9	1.38	1.83	2.26	2.82	3.25	4.30	4.78
	10	1.37	1.81	2.23	2.76	3.17	4.14	4.59
	11	1.36	1.80	2.20	2.72	3.11	4.03	4.44
	12	1.36	1.78	2.18	2.68	3.06	3.93	4.32
	13	1.35	1.77	2.16	2.65	3.01	3.85	4.22
	14	1.35	1.76	2.15	2.62	2.98	3.79	4.14
	15	1.34	1.75	2.13	2.60	2.95	3.73	4.07
	16	1.34	1.75	2.12	2.58	2.92	3.69	4.02
Degrees	17	1.33	1.74	2.11	2.57	2.90	3.65	3.97
of	18	1.33	1.73	2.10	2.55	2.88	3.61	3.92
Freedom	19	1.33	1.73	2.09	2.54	2.86	3.58	3.88
	20	1.33	1.73	2.09	2.53	2.85	3.55	3.85
	21	1.32	1.72	2.08	2.52	2.83	3.53	3.82
	22	1.32	1.72	2.07	2.51	2.82	3.51	3.79
	23	1.32	1.71	2.07	2.50	2.81	3.49	3.77
	24	1.32	1.71	2.06	2.49	2.80	3.47	3.75
	25	1.32	1.71	2.06	2.49	2.79	3.45	3.73
	26	1.32	1.71	2.06	2.48	2.78	3.44	3.71
	27	1.31	1.70	2.05	2.47	2.77	3.42	3.69
	28	1.31	1.70	2.05	2.47	2.76	3.41	3.67
	29	1.31	1.70	2.05	2.46	2.76	3.40	3.66
	30	1.31	1.70	2.04	2.46	2.75	3.39	3.65
	40	1.30	1.68	2.02	2.42	2.70	3.31	3.55
	60	1.30	1.67	2.00	2.39	2.66	3.23	3.46
	120	1.29	1.66	1.98	2.36	2.62	3.16	3.37

Table A.3

Table A.3 is taken from Table III of Fisher & Yates: STATISTICAL TABLES FOR BIOLOGICAL, AGRICULTURAL AND MEDICAL RESEARCH 6/e published by Addison Wesley Longman Ltd., (1963)

Chi-Square Distribution of Critical Values

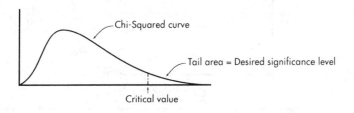

		Significance Level			
		.10	.05	.01	.001
	1	2.71	3.84	6.64	10.83
	2	4.61	5.99	9.21	13.82
	3	6.25	7.82	11.34	16.27
	4	7.78	9.49	13.28	18.47
	5	9.24	11.07	15.09	20.52
	6	10.64	12.59	16.81	22.46
	7	12.02	14.07	18.48	24.32
	8	13.36	15.51	20.09	26.12
Degrees	9	14.68	16.92	21.67	27.88
of	10	15.99	18.31	23.21	29.59
Freedom	11	17.28	19.68	24.72	31.26
	12	18.55	21.03	26.22	32.91
	13	19.81	22.36	27.69	34.53
	14	21.06	23.68	29.14	36.12
	15	22.31	25.00	30.58	37.70
	16	23.54	26.30	32.00	39.25
	17	24.77	27.59	33.41	40.79
	18	25.99	28.87	34.81	42.31
	19	27.20	30.14	36.19	43.82
	20	28.41	31.41	37.57	45.31

Table A.4

F Distribution of Critical Values

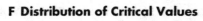

.05 Significance Level

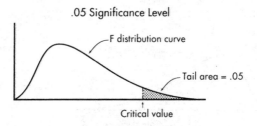

F distribution curve

Tail area = .05

Critical value

		Numerator Degrees of Freedom									
		1	2	3	4	5	6	7	8	9	10
Denominator Degrees of Freedom	1	161.40	199.50	215.70	224.60	230.20	234.00	236.80	238.90	240.50	241.90
	2	18.51	19.00	19.16	19.25	19.30	19.33	19.35	19.37	19.38	19.40
	3	10.13	9.55	9.28	9.12	9.01	8.94	8.89	8.85	8.81	8.79
	4	7.71	6.94	6.59	6.39	6.26	6.16	6.09	6.04	6.00	5.96
	5	6.61	5.79	5.41	5.19	5.05	4.95	4.88	4.82	4.77	4.74
	6	5.99	5.14	4.76	4.53	4.39	4.28	4.21	4.15	4.10	4.06
	7	5.59	4.74	4.35	4.12	3.97	3.87	3.79	3.73	3.68	3.64
	8	5.32	4.46	4.07	3.84	3.69	3.58	3.50	3.44	3.39	3.35
	9	5.12	4.26	3.86	3.63	3.48	3.37	3.29	3.23	3.18	3.14
	10	4.96	4.10	3.71	3.48	3.33	3.22	3.14	3.07	3.02	2.98
	11	4.84	3.98	3.59	3.36	3.20	3.09	3.01	2.95	2.90	2.85
	12	4.75	3.89	3.49	3.26	3.11	3.00	2.91	2.85	2.80	2.75
	13	4.67	3.81	3.41	3.18	3.03	2.92	2.83	2.77	2.71	2.67
	14	4.60	3.74	3.34	3.11	2.96	2.85	2.76	2.70	2.65	2.60
	15	4.54	3.68	3.29	3.06	2.90	2.79	2.71	2.64	2.59	2.54
	16	4.49	3.63	3.24	3.01	2.85	2.74	2.66	2.59	2.54	2.49
	17	4.45	3.59	3.20	2.96	2.81	2.70	2.61	2.55	2.49	2.45
	18	4.41	3.55	3.16	2.93	2.77	2.66	2.58	2.51	2.46	2.41
	19	4.38	3.52	3.13	2.90	2.74	2.63	2.54	2.48	2.42	2.38
	20	4.35	3.49	3.10	2.87	2.71	2.60	2.51	2.45	2.39	2.35
	21	4.32	3.47	3.07	2.84	2.68	2.57	2.49	2.42	2.37	2.32
	22	4.30	3.44	3.05	2.82	2.66	2.55	2.46	2.40	2.34	2.30
	23	4.28	3.42	3.03	2.80	2.64	2.53	2.44	2.37	2.32	2.27
	24	4.26	3.40	3.01	2.78	2.62	2.51	2.42	2.36	2.30	2.25
	25	4.24	3.39	2.99	2.76	2.60	2.49	2.40	2.34	2.28	2.24
	26	4.23	3.37	2.98	2.74	2.59	2.47	2.39	2.32	2.27	2.22
	27	4.21	3.35	2.96	2.73	2.57	2.46	2.37	2.31	2.25	2.20
	28	4.20	3.34	2.95	2.71	2.56	2.45	2.36	2.29	2.24	2.19
	29	4.18	3.33	2.93	2.70	2.55	2.43	2.35	2.28	2.22	2.18
	30	4.17	3.32	2.92	2.69	2.53	2.42	2.33	2.27	2.21	2.16
	40	4.08	3.23	2.84	2.61	2.45	2.34	2.25	2.18	2.12	2.08
	60	4.00	3.15	2.76	2.53	2.37	2.25	2.17	2.10	2.04	1.99
	120	3.92	3.07	2.68	2.45	2.29	2.17	2.09	2.02	1.96	1.91
	∞	3.84	3.00	2.60	2.37	2.21	2.10	2.01	1.94	1.88	1.83

Table A.5

Table A.5 is taken from Table V of Fisher & Yates: STATISTICAL TABLES FOR BIOLOGICAL, AGRICULTURAL AND MEDICAL RESEARCH 6/e published by Addison Wesley Longman Ltd., (1963)

Correlation Coefficients (r) Values

Table of the Correlation Coefficient

Degrees of Freedom	r				
	0.10	0.05	0.02	0.01	0.001
1	.988	.997	.999	1.000	1.000
2	.900	.950	.950	.990	.999
3	.805	.878	.934	.959	.992
4	.729	.811	.882	.917	.974
5	.669	.754	.833	.874	.951
6	.621	.707	.789	.834	.925
7	.582	.666	.750	.798	.898
8	.549	.632	.716	.765	.872
9	.521	.602	.685	.735	.847
10	.497	.576	.658	.708	.823
11	.476	.553	.634	.684	.801
12	.457	.532	.612	.661	.780
13	.441	.514	.592	.641	.760
14	.426	.497	.574	.623	.742
15	.412	.482	.558	.606	.725
16	.400	.468	.543	.590	.708
17	.389	.456	.528	.575	.693
18	.378	.444	.516	.561	.679
19	.369	.433	.503	.549	.665
20	.360	.423	.492	.537	.652
25	.323	.381	.445	.487	.597
30	.296	.349	.409	.449	.554
35	.275	.325	.381	.418	.519
40	.257	.304	.358	.393	.490
45	.243	.287	.338	.372	.465
50	.231	.273	.322	.354	.443
60	.211	.250	.295	.325	.408
70	.195	.232	.274	.302	.380
80	.183	.217	.256	.283	.357
90	.173	.205	.242	.267	.337
100	.164	.195	.230	.254	.321

Abridged from Table VI of "Statistical Tables for Biological, Agricultural and Medical Research." (R.A. Fisher and F. Yates : Oliver and Boyd).

Table A.6

Table A.6 is taken from Table VI of Fisher & Yates: STATISTICAL TABLES FOR BIOLOGICAL, AGRICULTURAL AND MEDICAL RESEARCH 6/e published by Addison Wesley Longman Ltd., (1963)

Constants and Formulas for Control Charts

Sub-group Size	$\bar{X}$ and R Charts				$\bar{X}$ and s Charts			
	Chart for Averages $(\bar{X})$	Chart for Ranges (R)			Chart for Averages $(\bar{X})$	Chart for Ranges Standard Deviations (s)		
	Factors for Control Limits	Divisors for Estimate of Standard Deviation	Factors for Control Limits		Factors for Control Limits	Divisors for Estimate of Standard Deviation	Factors for Control Limits	
n	A_2	d_2	D_3	D_4	A_3	c_4	B_3	B_4
2	1.880	1.128	-	3.267	2.659	0.7979	-	3.267
3	1.023	1.693	-	2.574	1.954	0.8862	-	2.568
4	0.729	2.059	-	2.282	1.628	0.9213	-	2.266
5	0.577	2.326	-	2.114	1.427	0.9400	-	2.089
6	0.483	2.534	-	2.004	1.287	0.9515	0.030	1.970
7	0.419	2.704	0.076	1.924	1.182	0.9594	0.118	1.882
8	0.373	2.847	0.136	1.864	1.099	0.9650	0.185	1.815
9	0.337	2.970	0.184	1.816	1.032	0.9693	0.239	1.761
10	0.308	3.078	0.223	1.777	0.975	0.9727	0.284	1.716
11	0.285	3.173	0.256	1.744	0.927	0.9754	0.321	1.679
12	0.266	3.258	0.283	1.717	0.886	0.9776	0.354	1.646
13	0.249	3.336	0.307	1.693	0.850	0.9794	0.382	1.618
14	0.235	3.407	0.328	1.672	0.817	0.9810	0.406	1.594
15	0.223	3.472	0.347	1.653	0.789	0.9823	0.428	1.572
16	0.212	3.532	0.363	1.637	0.763	0.9835	0.448	1.552
17	0.203	3.588	0.378	1.622	0.739	0.9845	0.466	1.534
18	0.194	3.640	0.391	1.608	0.718	0.9854	0.482	1.518
19	0.187	3.689	0.403	1.597	0.698	0.9862	0.497	1.503
20	0.180	3.735	0.415	1.585	0.680	0.9869	0.510	1.490
21	0.173	3.778	0.425	1.575	0.663	0.9876	0.523	1.477
22	0.167	3.819	0.434	1.566	0.647	0.9882	0.534	1.466
23	0.162	3.858	0.443	1.557	0.633	0.9887	0.545	1.455
24	0.157	3.895	0.451	1.548	0.619	0.9892	0.555	1.445
25	0.153	3.931	0.459	1.541	0.606	0.9896	0.565	1.435

$$UCL_{\bar{X}}, LCL_{\bar{X}} = \bar{\bar{X}} \pm A_2\bar{R} \qquad UCL_{\bar{X}}, LCL_{\bar{X}} = \bar{\bar{X}} \pm A_3\bar{s}$$

$$UCL_{\bar{R}} = D_4R \qquad\qquad UCL_{\bar{s}} = B_4s$$

$$LCL_{\bar{R}} = D_3R \qquad\qquad LCL_{\bar{s}} = D_3s$$

$$\hat{\sigma} = \bar{R}/d_2 \qquad\qquad\quad \hat{\sigma} = \bar{s}/c_4$$

Table A.7

Table of Random Numbers

```
07  28  68  61  81  38  11  98  34  74  64  03  48  09  18  10  15  25  98  80
29  24  86  11  41  21  16  12  96  17  56  61  49  32  48  35  43  29  34  12
76  05  58  54  35  55  35  59  07  19  00  92  65  95  34  88  26  32  61  36
95  01  20  28  66  31  15  92  14  33  39  98  55  85  71  35  82  04  51  64
73  89  25  53  83  33  75  79  98  20  09  06  76  92  43  42  55  86  41  67
41  58  46  41  68  72  73  78  34  65  87  08  10  93  46  00  32  48  29  68
53  46  33  57  86  99  47  87  14  55  98  93  72  15  77  23  13  26  37  20
39  46  65  77  16  92  33  65  57  49  18  41  87  68  05  23  73  33  55  49
40  98  58  06  54  13  55  31  86  05  34  94  43  59  08  54  86  44  59  84
06  45  65  80  97  46  95  38  82  01  88  12  28  75  93  39  33  60  00  48
84  72  36  35  94  11  36  23  17  09  95  90  26  46  90  70  81  40  77  38
61  14  68  60  77  44  75  28  56  67  36  58  03  82  16  76  39  12  73  70
07  47  15  19  64  62  17  97  36  08  22  55  58  81  17  77  83  65  75  05
70  43  84  46  41  98  44  54  23  72  39  79  53  16  88  04  66  00  66  43
57  10  02  26  17  12  56  48  43  97  65  06  21  97  65  97  95  77  93  01
95  01  58  34  51  77  89  80  79  72  60  94  43  05  89  83  88  15  09  58
53  00  18  66  58  39  02  95  62  79  35  52  01  06  50  18  98  88  87  81
51  86  20  34  89  54  54  61  15  00  96  89  11  34  05  18  26  77  17  23
38  63  42  41  87  99  37  18  91  08  55  42  27  51  69  48  94  14  70  96
47  77  39  28  14  56  98  96  73  22  31  67  20  90  85  04  01  87  42  17
26  20  46  66  36  28  98  66  97  56  78  29  19  53  46  08  20  30  55  61
58  58  28  68  36  45  83  66  12  05  17  37  74  90  81  86  99  04  17  90
80  83  75  20  32  63  09  41  69  12  43  82  63  40  08  89  71  89  68  44
40  90  05  68  85  00  90  91  49  16  23  00  26  56  52  66  71  22  63  40
77  38  50  26  29  57  56  31  37  52  88  88  37  72  14  52  73  79  23  79
51  62  77  67  70  21  17  88  22  26  66  77  78  55  87  14  39  07  31  67
66  81  52  18  87  47  01  60  71  73  90  72  90  39  37  64  44  26  82  07
67  72  78  24  07  12  61  67  78  85  92  68  95  24  69  57  74  13  28  64
14  29  00  91  50  43  64  63  85  17  54  46  92  58  58  52  97  54  84  09
30  89  99  07  56  26  49  27  83  67  52  35  36  93  63  60  15  71  16  34
26  42  43  27  81  79  67  35  84  28  64  59  79  16  11  54  85  34  01  49
98  05  34  47  71  14  87  98  70  21  53  51  01  46  60  71  19  33  62  43
02  82  10  42  11  62  87  83  16  96  34  46  04  25  33  69  55  37  82  29
99  88  34  85  46  77  12  00  89  17  04  48  85  62  32  77  08  24  88  65
83  59  57  38  84  22  08  75  21  10  58  75  87  70  19  07  94  83  09  37
76  27  52  23  67  14  39  88  57  00  72  71  21  68  81  49  24  94  19  37
03  80  24  56  17  64  66  90  80  09  62  03  65  61  66  39  83  87  41  95
40  86  98  74  63  72  14  00  08  38  25  25  37  93  89  96  74  66  36  06
38  02  78  20  39  15  04  67  68  27  46  22  43  79  26  45  45  17  66  13
19  51  85  12  56  95  63  15  44  74  88  26  02  10  68  09  84  86  26  81
```

Table A.8

Appendix B

Analysis of Variance Procedure

1. Square the individuals and add.

2. Obtain the total of each factor, square this total, sum these squares, and divide this total by the number of individuals in each factor.

3. Obtain the total of each interaction (of concern), square these totals, sum these squares, and divide this total by the number of individuals in each interaction.

4. Obtain the grand total of all individuals, square this grand total, and divide the square by all individual observations. This statistic is often referred to as the correction factor.

5. Obtain the total variance, this is the total sum of squares minus the correction factor. It is the value in step minus the value from step 4:

$$s^2 \text{ total} = x^2 - \frac{(x)^2}{n}$$

6. Main effect–Sum-of-Square calculation–Value in step 2 minus the correction factor from step 4.

7. 1st Order Interaction–Sum-of-Square calculation–Value from step 3 minus the correction factor from step 4, minus the main effect from step 6.

8. Residual–Sum-of-Square calculation–Total from step 5 minus the main effects from step 6 minus the any interactions from step 7. Degrees of Freedom–$(n - 1)$ for each factor.

10. Calculate the F Ratios for each factor–Factor Variance (Residual Variance.

Test of Significance for each factor: compare the F Ratios from step 10 and their respective Degrees of Freedom with the Values in the F Ration Tables. See table A.5 in Appendix A.

Appendix C

Other Factorial Designs

A factorial design involves n factors where the combinations of any one version with the versions of the others appear only once.

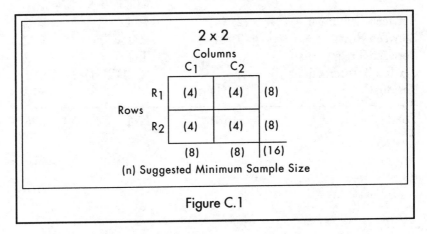

Figure C.1

Analysis of Variance Table

Source of Variance	Sum of Squares	Degrees of Freedom	
Between Columns	(2)-(4)	C-1	= 1
Between Rows	(3)-(4)	R-1	= 1
C x R Interactions		(C-1)(R-1)	= 1
Residual	(5)-(1)		= 12
Total	(5)-(4)	N-1	=15

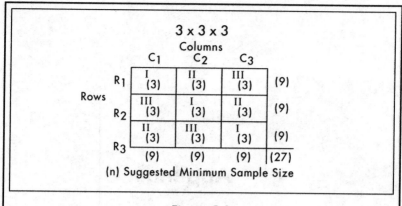

Figure C.2

Analysis of Variance Table

Source of Variance	Sum of Squares	Degrees of Freedom	
Between Columns	(2)-(4)	C-1	= 2
Between Rows	(3)-(4)	R-1	= 2
Between Treatments		T-1	= 2
C x R x T Interactions		(C-1)(R-1)(T-1)	= 1
Residual	(5)-(1)		=12
Total	(5)-(4)	N-1	=26

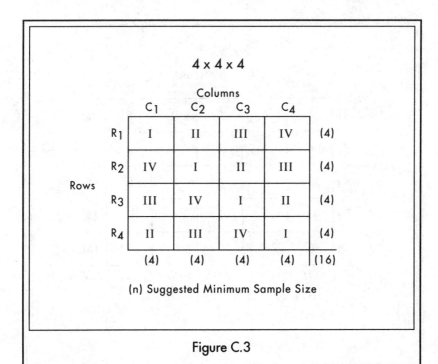

Figure C.3

Analysis of Variance Table

Source of Variance	Sum of Squares	Degrees of Freedom	
Between Columns	(2)-(4)	C-1	= 1
Between Rows	(3)-(4)	R-1	= 1
C x R Interactions	(3)-(4)	T-1	= 1
Residual	(5)-(1)		=12
Total	(5)-(4)		=15

5 x 5 x 5

Factor 1 (Columns)

		C_1	C_2	C_3	C_4	C_5	
	R_1	A	B	C	D	E	(5)
	R_2	B	A	E	C	D	(5)
Factor 2 (Rows)	R_3	C	D	A	E	B	(5)
	R_4	D	E	B	A	C	(5)
	R_5	E	C	D	B	A	(5)
		(5)	(5)	(5)	(5)	(5)	(25)

(n) Suggested Minimum Sample Size

Figure C.4

Analysis of Variance Table

Source of Variance	Sum of Squares	Degrees of Freedom	
Between Columns	(2)-(4)	C-1	= 4
Between Rows	(3)-(4)	R-1	= 4
C x R Interactions	(3)-(4)	T-1	= 4
Residual	(5)-(1)		=12
Total	(5)-(4)		=24

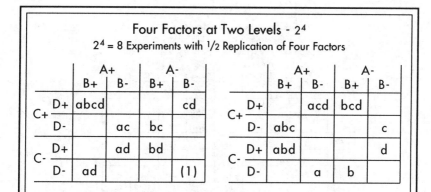

Figure C.5

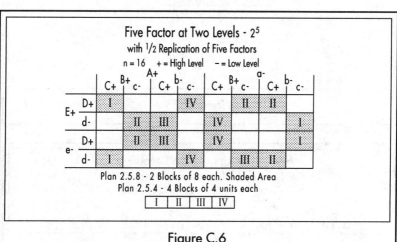

Figure C.6

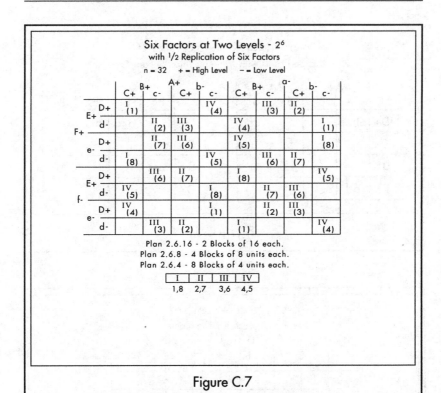

Figure C.7

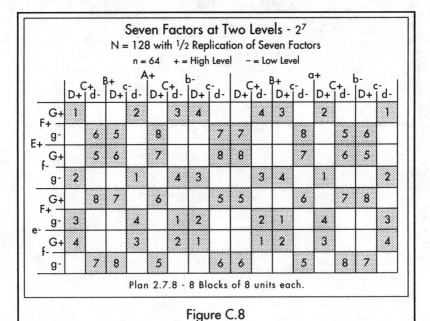

Figure C.8

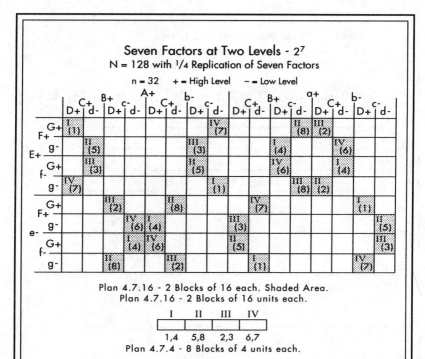

Plan 4.7.16 - 2 Blocks of 16 each. Shaded Area.
Plan 4.7.16 - 2 Blocks of 16 units each.

I	II	III	IV
1,4	5,8	2,3	6,7

Plan 4.7.4 - 8 Blocks of 4 units each.

1	2	3	4	5	6	7	8

Figure C.9

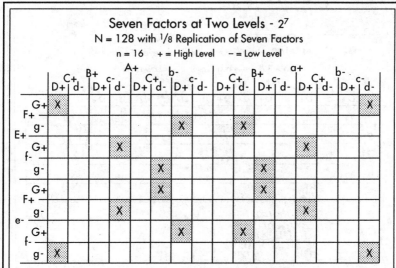

Plan 8.7.8 - 2 Blocks of 8 units each. Shaded Area.
Plan 8.7.4 - 4 Blocks of 4 units each.

Figure C.10

Appendix D

Use of Statistical Calculators

Inexpensive hand-held calculators are now available that will solve statistical equations with no difficulty. It is not necessary to have a math background beyond simple high school algebra to use them. They come with a self-instructional booklet that is easy to use and is "user oriented." Data values are entered into the calculator register. Pressing a statistical function key then analyzes the data. To use the calculator:

1. First clear the statistical registers. [CE/C] [CL X]

2. To enter statistical data values press the numerical value keys [O through 9] or a combination of number keys, for example [8] + [9] for [89].

3. The $x^2 - \dfrac{(x)^2}{n}$ key enters the values into the statistical registers.

4. The [X] adds the numbers that have seen entered and divides the numbers by the "n" number of values entered to give the average $(\overline{X})$

5. The [s] or $[\sigma_{n-1}]$ calculates the sampling standard deviation of the values entered by using the following formula:

$$s = \sqrt{\dfrac{\Sigma X^2 - \dfrac{(\Sigma X)^2}{n}}{n-1}}$$

6. Some calculators calculate the Z values for the area under the curve.

The hand-held calculator, in addition to calculating X, s, and Z, may also be used to find the following mathematical functions from the key board or a register.

$\left[\sigma^n\right]$ = standard deviation of population data.

$\left[\Sigma X +\right]$ = enters the data points for calculations.

$\left[\bar{y}\right]$ = mean of the y values.

$\left[\Sigma X\right]$ = sum of the X values.

$\left[\Sigma X^2\right]$ = sum of squares of the x data

$\left[R(t)\right]$ = computes the Z value.

$\left[\sigma^{n-1}\right]$ = standard deviation of sample data.

$\left[e^x\right]$ = calculates e to the xth power.

$\left[\sqrt{x}\right]$ = calculates the square root of the displayed number.

$\left[1/X\right]$ = calculates the reciprocal of the displayed number.

$\left[y^x\right]$ = calculates y to the xth power.

$\left[X!\right]$ = calculates the factorial of a displayed number.

$\left[EXP\right]$ = enables you to enter a number with an exponent.

$\left[SUM\right]$ = adds a number in the display to a number in memory.

$\left[\log\right]$ = calculates the natural logarithm of a displayed number.

$\left[Rnd\right]$ = generates a random number.

$\left[nPr\right]$ = calculates the number of permutations of n items.

$\left[nCr\right]$ = calculates the number of n item combinations taken r at a time.

Appendix E

$\overline{X}$ and R Charts

For $\overline{X}$ and R charts, it is necessary to have measurements that are capable of showing actual degrees of variation. Measurements of this type are called variables measurements. Common examples of variables measurements are length in feet, diameter in inches, resistance in ohms, noise in decibels, weld strength in grams or pounds, etc. Many characteristics which are not ordinarily measured in the shop (cracks, appearance, or workmanship, for example) can be put in a form which is roughly equivalent to variables measurements, if necessary.

B-3.2 Definitions and symbols for $\overline{X}$ and R charts
The following symbols are used in connection with $\overline{X}$ and R charts:

- X = an individual reading or observation.
- n = the number of observations in a group or set, often referred to as the sample size. "n" may be 2, 3, 4, 5 or more, but should not be greater than 10.
- $\overline{X}$ (X bar) = the average of a group of Xs. $\overline{X}$ is calculated by the formula

$$\overline{X} = \frac{X_1 + X_2 + X_3...X_n}{n}$$

- $\overline{\overline{X}}$ (X double bar) = the average of a series of $\overline{X}$ values.

- R (range) = the difference between the largest and smallest reading in a sample of n values.
- $\overline{R}$ (R bar) = the average of a series of R values.
- A_2 = a factor used in calculating the control limits for the $\overline{X}$ chart.
- D_4 = a factor used in calculating the lower control limit for the R-chart.
- D_3 = a factor used in calculating the lower control limit for the R-chart.

The factors A_2, D_4, and D_3 vary with the size of the sample.

Table of Factors for $\overline{X}$ Bar and R Charts				
Number of Observations in Sample	$\overline{X}$ Chart Factors for Control Limits	R Chart Factors for Control Limits		Divisors for Estimate of Standard Deviation
(n)	(A_2)	Upper (D_3)	Lower (D_4)	(d_3)
2	1.880	0	3.267	1.128
3	1.023	0	2.574	1.693
4	.729	0	2.282	2.059
5	.577	0	2.114	2.326
6	.483	0	2.004	2.534
7	.419	.076	1.924	2.704
8	.373	.136	1.864	2.847
9	.337	.184	1.816	2.970
10	.308	.223	1.777	3.078
$\overline{\overline{X}} \pm A_2 \overline{R}$		LCL = $D_3\overline{R}$	UCL = $D_3\overline{R}$	$\sigma = \overline{R}/d_2$

Table E.1

B-3.3 Directions for making an $\overline{X}$ and R chart

Always make the R-chart before the $\overline{X}$-chart. Proceed as follows.

R-chart

(1) Decide on the sample size (n) to be used.

(2) Obtain a series of groups of measurements, each group containing "n" measurements. Have 20 or more groups, if possible, but not less than 10 groups.

(3) Compute R for each sample and take the average of the Rs ($\overline{R}$). This is the centerline for the R chart. It is drawn as a solid horizontal line. See figure E.1.

(4) Multiply $\overline{R}$ by D_4 and D_3 to obtain upper and lower control limits for the R chart. The control limits are drawn as dotted horizontal lines. See figure E.1.

(5) Obtain a piece of graph paper (or a standard control chart form) and set up an appropriate scale. Be careful not to make the R-chart too wide. Set up the R-chart at the bottom of the sheet as shown in figure E.1.

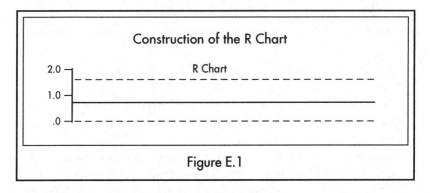

Figure E.1

(6) Plot on this chart the successive values of R and connect the points with straight lines.

(7) Mark Xs on the chart, if necessary, in accordance with the rules given Appendix H.

$\overline{X}$-chart

(1) Use exactly the same groups of measurements that were used for the R chart.

(2) Calculate $\overline{X}$ for each sample and take the average of the Xs. This is the centerline for the $\overline{X}$ chart. It is drawn as a solid horizontal line. See figure E.2.

(3) Multiply $\overline{R}$ by A_2 to get the width of the control limits for the $\overline{X}$ chart. Add the $A_2\overline{R}$ value to (and subtract it from) $\overline{X}$ to get the location of the control limits.

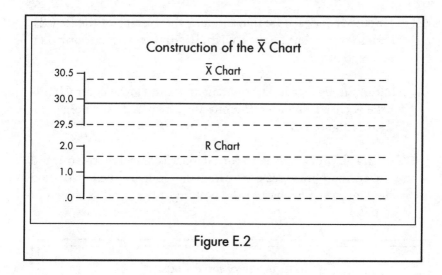

Figure E.2

Upper control limit for $\overline{X}$-chart $= X + A_2 \overline{R}$.
Lower control limit for $\overline{X}$-chart $= X - A_2 \overline{R}$.
The control limits are shown as dotted horizontal lines. (See fig. E.2.)

(4) Choose a scale for the $\overline{X}$-chart that is properly related to the scale already chosen for the R-chart. The scale should be such that the distance between the control limts on the $\overline{X}$-chart are roughly similar to the distance between the control limits on the R chart. This relationship can be obtained as follows:
   a. For samples of 2, use the same spacing on the $\overline{X}$ chart as on the R chart.
   b. For samples of 5, let each division on the graph paper represent an increment of measurement half as large as on the R chart.
   c. For samples of 10, let each division on the graph paper represent an increment of measurement one-third as large as on the R chart.
This relationship between the scales corresponds roughly to $\frac{1}{\sqrt{n}}$. For other sample sizes, use the scale nearest to those given above.

(5) Set up an $\overline{X}$-chart in the upper portion of the sheet as shown in figure E.2.

(6) Plot on this chart the successive values of $\overline{X}$ and connect the points with straight lines.

(7) Mark Xs on the chart, if necessary, in accordance with the rules given in Appendix H.

(8) If drawing limits are specified for the characteristic being plotted, draw arrows in the left-hand margin along the X scale to represent the drawing limits.

(9) Interpret the chart as explained in Appendix H.

Directions for samples of 5
Since samples of 5 are used more commonly than any others, the instructions for this sample size are summarized for easy reference in figure E.3.

$\overline{X}$ *and R Chart for n = 5*
Divide the data into groups of 5.
R- chart
For each group obtain R.
Centerline on R chart = $\overline{R}$.
Lower Control Limit = 0.
Upper control limit = 2.11 $\overline{R}$.

$\overline{X}$*-Chart*
For each group obtain $\overline{X}$.
Centerline on $\overline{X}$ chart = $\overline{\overline{X}}$.
Control limits = $\overline{\overline{X}} \pm .58\overline{R}$.

Figure E.3

B-3.4. Example of calculations for an $\overline{X}$ and R chart
Obtain a set of data as shown in figure E.4.

(1) Centerline for R-chart:

$$\overline{R} = \frac{Total\ of\ Rs}{No.\ of\ Samples} = \frac{31.8}{20} = 1.59$$

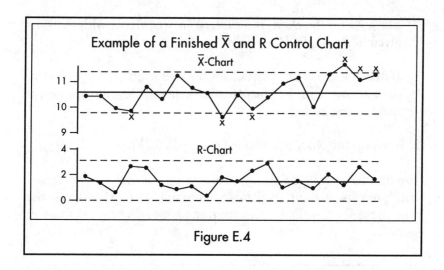

Figure E.4

(2) Control limits for R-chart:

$$D_4 \overline{R} = (2.11)(1.59) = 3.35$$
$$D_3 \overline{R} = (0)(1.59) = 0$$

(3) Centerline for $\overline{X}$-chart:

$$\overline{X} = \frac{Total\ of\ Xs}{No.\ of\ Samples} = \frac{213.20}{20} = 10.66$$

(4) Control limits for R-chart:

$$\overline{X} + A_2 R = 10.66 + (.58)(1.59)$$
$$= 10.66 + .92 = 11.58$$
$$\overline{X} - A_2 R = 10.66 - .92 = 9.74$$

The completed chart is shown in figure E.4.

Appendix F

p-Charts

B-4.2. Definitions and symbols for a p-chart
The following symbols are used in coneection with p-charts:

n = number of units in a sample

p = fraction defective in a sample

$$= \frac{total\ number\ of\ defective\ units\ in\ a\ sample}{sample\ size}$$

$\bar{p}$ = average fraction defective in a series of samples

$$= \frac{total\ no.\ defective\ units\ in\ all\ samples\ in\ the\ series}{total\ nos.\ checked\ in\ all\ samples}$$

B-4.3 Directions for making a p-chart
Proceed as follows:

(1) Obtain a series of samples of some appropriate size. Convenient sample sizes are 50 and 100. The "sample" may actually be the complete lot if the entire lot has been checked. Have 20 or more groups if possible, but not less than 10 groups.

(2) Count the number of defective units (warped, undersize, oversize, or whatever the characteristic may be in which you are interested). Calculate the value of p for each sample.

(3) Calculate $\bar{p}$ (the average fraction defective) as shown in B-4.2. This is the centerline for the p-chart. It is drawn as a solid horizontal line. See figure F.1.

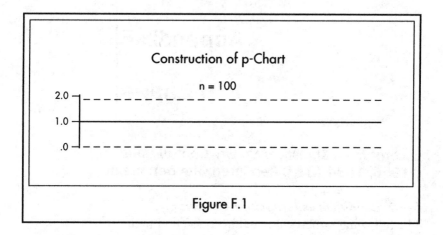

Figure F.1

(4) Calculate upper and lower control limits for the p-chart by using the following formulas.

Upper control limit for p-chart =

$$\bar{p} + 3\sqrt{\frac{\bar{p}(1-\bar{p})}{n}}$$

Lower control limit for p-chart =

$$\bar{p} - 3\sqrt{\frac{\bar{p}(1-\bar{p})}{n}}$$

The control limits are drawn as dotted horizontal lines. See figure F.2.

(5) Decide on an appropriate scale and set up a p-chart on graph paper or a standard control chart form. Be careful not to make the p-chart too wide. See figure F.2.

(6) Plot on this chart the successive values of p and connect the points with straight lines.

(7) Mark Xs on the chart, if necessary, in accordance with the rules given in Appendix H.

Average value of n
If the samples to be used for a p-chart are not the same size, it is sometimes permissible to use the average sample size for the series in calculating the control limits. The largest sample in the series should not be more than twice the average sample size, and the smallest sample in the series should not be less than half the average sample size.

If the individual samples vary more than this, either separate or combine samples to make them of suitable size, or calculate control limits separately for the samples which are too large or too small.

p-charts are most useful when this problem is avoided by keeping the sample size constant.

B-4.4 Example of calculations for a p-chart
Obtain a set of data as shown in table F.1.

Typical Data for a p-Chart			
Date	No. in Sample	No. Defective Found in Sample	% Defective in Sample
11/2	1517	70	4.61
11/3	1496	91	6.08
11/4	1488	94	6.32
11/5	1696	37	2.18
11/6	1427	58	4.06
11/9	1723	118	6.85
11/10	1524	88	5.78
11/11	1371	55	4.01
11/12	1517	159	10.48
11/13	1488	94	6.32
Total	15,247	864	5.67

Table F.1

(1) Centerline for p-chart:

$$\bar{p} = \frac{\textit{total no. defective units in all samples in the series}}{\textit{total nos. checked in all samples}}$$

$$= \frac{864}{15247} = .0567(5.12)$$

(2) Average n for p-chart:

$$\text{average n} = \frac{\textit{total number checked in all samples}}{\textit{number of all samples in the series}}$$

$$= \frac{15247}{10}$$

$$= 1525$$

(3) Upper control limit for p-chart:

$$= \bar{p} + 3\sqrt{\frac{\bar{p}(1-\bar{p})}{n}}$$

$$= .0567 + 3\sqrt{\frac{(.0567)(.9433)}{1525}}$$

$$= .0567 + 3\sqrt{\frac{.0534}{1525}}$$

$$= .0567 + (3)(.00592)$$

$$= .0567 + .0178$$

$$= .0745$$

$$= 7.45\%$$

Lower control limit for p-chart:

$$= .0583 - .0186 = .0397$$

These may be converted to percentages by multiplying by 100. The completed chart is shown in figure F.2.

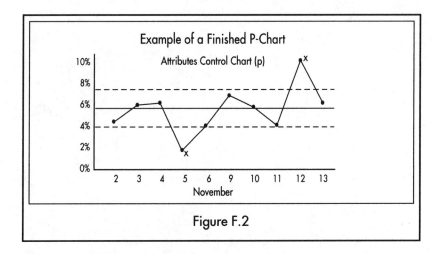

Figure F.2

B-4.5 Stairstep limits on a p-chart

The control limits in figure E.4 were calculated by using an average value of n. It is also possible to calculate control limits for each point on a p-chart individually, using the individual values of n. When this is done, the control limits may have an irregular "stairstep" effect. Where the sample size is small, the control limits are wider. Where the sample size is large, the control limits are tighter. The pattern is marked with Xs in the same way as if they were ordinary control limits.

"Stairstep" limits make a p-chart more difficult to interpret, and their use should be avoided. As mentioned above, limits of this kind are occasionally used for individual samples, where the sample is too small or too large to be covered by the "average value of n."

Appendix G

Charts with Moving Range Limits

**Excerpt from *Statistical Quality Control Handbook*
©1956, 1984 AT&T. Reprinted with permission.**

To plot this type of chart, proceed as follows:

(1) Start with a series of individual numbers. Have 20 or more numbers if possible, but not less than 10 numbers.

(2) Take the difference between the first and second numbers, and record it; then the difference between the second and third numbers, etc. Continue in this way until you have taken the difference between the next-to-last and the last numbers. The number of differences or "ranges" should be one less than the number of individuals in the series. The differences are calculated without regard to sign. That is, it does not matter whether they are plus or minus (gain or loss).

(3) Take the average of the original numbers in the series ($\overline{X}$). This is the centerline for the chart. It is drawn as a solid horizontal line.

(4) Take the average of the "ranges" obtained in step (2). Be sure to divide by the number of ranges, which is one less than the number of original measurements. This average range is called $M\overline{R}$.

(5) Multiply $M\overline{R}$ by 2.66 (a constant factor) to get the width of the control limits for the moving range chart. Add this value to (and subtract it from) $\overline{X}$ to get the location of the control limits.

$$Control\,limits = X \pm 2.66 M\overline{R}$$

The control limits are shown as dotted horizontal lines. (See fig. G.1.)

(6) Set up a chart on graph paper or standard control chart form. Be careful not to make the chart too wide. See figure G.1.

(7) Plot on this chart the series of original numbers, and connect the points with straight lines. Do not plot the moving ranges calculated in Step (2).

(8) Mark Xs on the chart, if necessary, in accordance with the rules given in Appendix H.

B-5.4. Example of calculations for a chart with "moving range" limits
Obtain a set of data as shown in table G.1.

(1) Centerline for chart:

$$\overline{X} = \frac{total\,of\,individuals}{number\,of\,individuals}$$

$$= \frac{460.4}{15} = 30.7$$

(2) Average moving range:

$$M\overline{R} = \frac{total\,of\,moving\,ranges}{number\,of\,moving\,ranges}$$

$$= \frac{53.6}{14} = 3.8$$

(3) Upper control limit:

$$= \overline{X} + 2.66(M\overline{R})$$

$$= 30.7 + [(2.66)3.8]$$

$$= 30.7 + 10.1$$

$$= 40.8$$

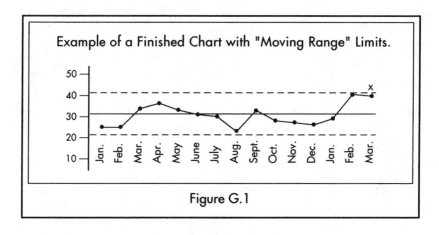

Figure G.1

Typical Data for a Chart with "Moving Range" Limits.

Earnings of a Group of Workers	% Earned	Change from Preceding Month (Moving Range)
January (last year)	25.0	. . .
February	25.3	.3
March	33.8	8.5
April	36.4	2.6
May	32.2	4.2
June	30.8	1.4
July	30.0	.8
August	23.6	6.4
September	32.3	8.7
October	28.1	4.2
November	27.0	1.1
December	26.1	.9
January (this year)	29.1	3.0
February	40.1	11.0
March	40.6	.5
	460.4	53.6

Table G.1

(4) Lower control limit

$$= X - 2.66(M\overline{R})$$
$$= 30.7 - 10.1$$
$$= 20.6$$

The completed chart is shown in figure G.1.

Appendix H

Tests for
Unnatural Patterns

Excerpt from *Statistical Quality Control Handbook*
©1956, 1984 AT&T. Reprinted with permission.

The most important of the tests for unnatural patterns are the tests for "instability." These are tests to determine whether the cause system is changing. In applying these tests consider only the area between the centerline and one of the control limits. Divide this area mentally into three equal zones.

Since the control limits are 3 sigma limits, each of the zones is one sigma in width. For this reason, the zones are sometimes referred to as the "one sigma zone," the "two sigma zone," etc. See figure H.1.

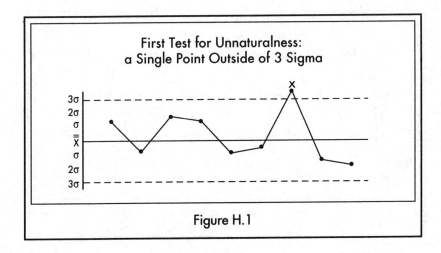

Figure H.1

The pattern is unnatural if any of the following combinations are formed in the various zones:

Test 1. A single point falls outside of the 3 sigma limit (beyond Zone A). See figure H.2.

Mark the unnatural point with an "x."

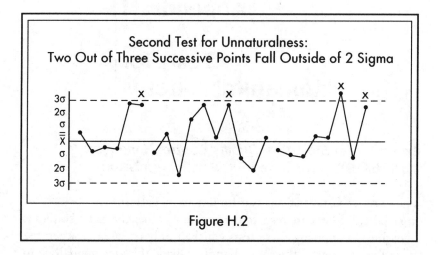

Figure H.2

Test 2. Two out of three successive points fall in Zone A or beyond. (Note: The odd point may be anywhere. Only the two points count.) See figure H.3.

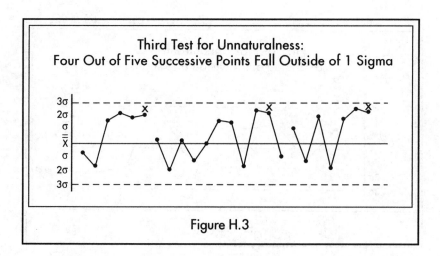

Figure H.3

Mark only the second of the two points with an "x," since the second point is necessary to produce a reaction to the test. In the last example above, the point which is third from the end is marked because it reacted to Test 1, and not because it was part of the test for "2 out of 3."

Test 3. Four out of five successive points fall in Zone B or beyond. (Note the odd point may be anywhere. Only the four points count.) See Figure H.4.

Mark only the last of the four points with an "x," since there is no reaction to the test until the fourth point.

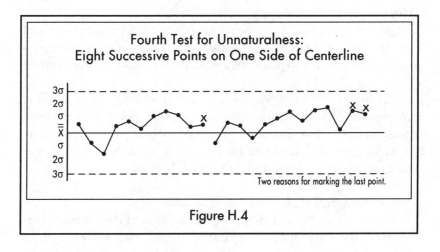

Figure H.4

Test 4. Eight successive points fall in Zone C or beyond. (This is sometimes expressed as "eight points in a row on one side of the centerline.")

Mark only the eight point with an "x," since all eight points are necessary to produce a reaction to the test.

In applying the tests, start with any point you choose (usually the last plotted point) and count backward as many points as are required to make the test. In the second example in figure H.4, the first "x" was arrived at by starting with the twelfth point and counting back to the fifth.

It is possible for the same point to react to more than one test. For example, in the last portion of figure H.4, the final point reacts to the test for "8 in a row" and also the test for "4 out of 5." In this case there are two reasons for marking the point with an "x." Do not, however, show more than one "x" for the point.

Marking the Xs
In marking the Xs, always put the "x" a uniform distance from the point being marked (preferably about 1/8 inch). Put it directly above the point if the point is in the upper half of the control chart, and directly below the point if the point is in the lower half of the control chart. That is, put the "x" on the side that is away from the centerline.

Interpretation of the Xs
The greater the instability in the system of causes, the more points will tend to react to these tests and be marked with Xs. After the pattern is marked, it is possible to judge the amount of instability by the number of Xs.

In looking for the *causes* which are producing the instability, remember that the causes may have affected more points than the ones actually marked. If a point has been marked for being the eighth on one side of the centerline, the cause has probably been in the picture for the whole run of eight points, and quite possibly before.

Applying the tests to the opposite half of the chart
The same tests for instability apply to both halves of the control chart. However, they are applied separately to the two halves, not in combination. For example, two points do not count in the "2 out of 3" if one is in Zone A on the upper half of the chart and the other is in Zone A on the lower half of the chart. Both of the points that count must be in the same half of the chart.

Appendix I

Relationship between $\overline{X}$-Chart and Specification

**Excerpt from *Statistical Quality Control Handbook*
©1956, 1984 AT&T. Reprinted with permission.**

To find the relationship between the process and the specification proceed as follows. *Both the $\overline{X}$-Chart and the R-chart must be in control before this check is made.*

(1) In the left-hand margin of the chart, along the $\overline{X}$ scale, draw one or more arrows to represent the specification limits. See figure I.1. The arrows may represent the tolerances specified on a drawing, a proposed specification, or merely some standard we wish to hold for economic reasons. Label each arrow "Maximum" or "Minimum," depending on whether it is an upper or lower limit.

(2) Check the shape of the distribution roughly by plotting the data in a frequency distribution. If the shape is roughly symmetrical, make the calculation explained in Step (3). If the shape is noticeably skewed, with a long thin tail on one side and a sharp cut-off on the other, proceed as in Step (4).

(3) For a symmetrical distribution, note the distance between the centerline on the $\overline{X}$ chart and one of the control limits. Multiply this distance by n and note whether it falls outside of the specification. If it does, some of the product from which these samples were taken is probably out of limits.

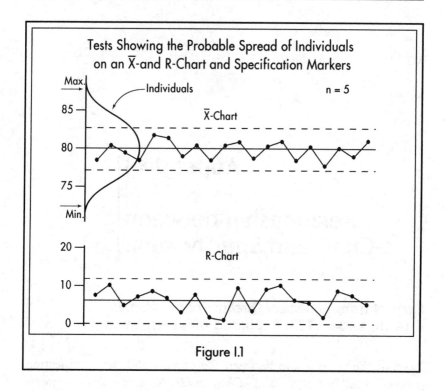

Figure I.1

(4) For unsymmetrical distributions make a similar calculation, but allow more on the long side and less on the short side.

In the case of samples of 5, it is convenient to memorize the following:

    a. If the distribution is symmetrical, the space between the control limit and the arrow should not be less than the space between the control limit and the centerline.

    b. If the arrow is opposite the centerline, about 50% of the product is probably out of limits.

    c. If the arrow is opposite the centerline, about 50% of the product is probably out of limits.

It is sometimes helpful to sketch the distribution of individuals on the $\overline{X}$ chart as shown in figure I.1. This provides a quick visual means of judging whether there is a conflict between the process and the specification.

Symbols and Acronyms

A_2 A multiplier of R in the construction of the control limits for an $\overline{X}$-chart.

A_3 A multiplier of s in the construction of the control limits for an $\overline{X}$-chart.

μ Alpha. Level of significance–risk of an error of the first kind.

b Beta. Risk of an error of the second kind.

$b_3 b_4$ Multipliers of s used to calculate the lower and upper control limits for sample standard deviations.

c The number of nonconformities in a sample. Used in the construction of a c-chart.

c The average number of nonconformitites in samples of constant size n.

c_4 A divisor of R used to estimate the process standard deviation.

Cpk The process capabilitiy index.

d_2 A divisor of R used to estimate the process standard deviation.

DF Degrees of freedom.

D_3D_4 Multipliers of R used to calculate the lower and upper control limits for ranges.

e   The constant 2.718.

EVOP   Evolutionary operation.

F   The symbol F is often identified with class frequency, Fisher's Ratio, F-test Statistic, and F-distribution.

H_0 Null hypothesis.

i   The class interval of a frequency distribution, used as a subscript to differentiate a variable.

k   Used as a subscript to differentiate a quantity, such as the number of subgroups being used to calculate control limits.

LCL   Lower control limit

LSL   Lower specification limit

M.S.   Mean square

μ (Mu.) Signifies population average.

N   Lot size.

n   Sample size.

np   The number of nonconforming units in a sample size n. Number of subgroups being used to calculate control limits.

p   Proportion or percent.

R   Subgroup range. Highest minus lowest value.

r   Correlation coefficient.

s (σ)   Sample standard deviation. (Standard deviation.)

s^2 Sample variance.

Σ Sigma, upper case. Signifies "sum of."

S.S.   Sum of squares.

t t   Test statistic.

u   Count per unit.

UCL   Upper control limit.

USL   Upper specification limit.

σ^2 Sigma squared. Variance. Population variance.

X   Observed value. Independent variable. X axis.

Y   Observed value. Dependent variable. Y axis.

Z   A variable measured from the population mean in terms of the population standard deviation.

χ^2 Chi-Square test statistic.

Glossary

accuracy. The closeness of agreement between an observed value and an accepted reference value.

alpha (α). The probability percentage of making a wrong decision.

analysis of variance. A technique for analyzing the total variation of a set of data into meaningful component parts associated with specific sources of variation for the purpose of estimating variance components.

assignable cause. Object of an analysis to detect variations attributable to special causes.

asterisks (or Xs). A mark on a control chart to identify an assignable cause.

attribute data. A measurement of quality consisting of noting the presence (or absence) of some characteristic or attribute in each of the items in the group under consideration.

Bartletts test. A statistical test to determine whether a set of variances could all have come from the same universe.

batch. A definite quantity of some product or material produced under conditions that are considered uniform.

beta (β). The failure to reject a false hypothesis.

bias. A systematic error that contributes to the difference between a population mean of measurements and an accepted reference value.

bimodal. Two overlapping distributions.

binomial distribution. A discrete distribution that depends on only two parameters, p and N.

block. A subdivision of a designed experiment into a group of relatively homogenous experimental units between which the experimental error can be expected to be smaller than would be expected should a similar number of units be randomly located within the entire experimental space.

box-and-whisker plot. A graphical technique for summarizing one or more histograms.

brainstorming. A problem-solving technique that generates a great variety of ideas about a subject using a group of people.

c-chart. A special type of attributes control chart that uses the number of defects instead of the number of defectives.

capability. Refers to the normal behavior of a process when operating in a state of statistical control.

capability index. The ratio of the specification limits to the 6 sigma process limits.

cause-and-effect diagram. A skeleton diagram used for selecting the vital factors for analysis and their relationships.

center line. The line on a control chart that represents the average or median value of the items being plotted.

central limit theorem. When the sample size is large, probabilities involving the average can be computed exactly as described by the standard normal distribution.

characteristic. A property that helps differentiate between items on which data can be collected.

chi-square. A measure of the relationship used to determine whether the frequency with which a given event has occurred is significantly different from what was expected.

cluster. A group with similar properties. For control charts and scatter plots it is a group of points falling in the same area of the chart.

coding. Referring a variable to an arbitrary origin and possibly expressing it in some other unit(s).

common cause. A source of variation that affects all the individual values of the process output being studied; it appears as part of the random process variation.

confidence interval. The probability that an interval about a sample statistic has a designated chance of actually including the population parameter.

confidence limit. The end points of a confidence interval.

confounding. Combining indistinguishable effects of a factor with the effects of other factors.

control limit. A line (or lines) on a control chart used as a basis for judging significance of the variation from subgroup to subgroup. Variation beyond a control limit is evidence that special causes are affecting the process. The limits are calculated from process data and are not to be confused with process specifications.

contingency matrix. A table of rows and columns that contains the correlation coefficients of all combinations.

correlation coefficient. The ratio of explained variation to the total variation. The quantity r, called the coeeficient of correlation, varies between -1 and +1.

critical value. The Z values outside the range -1.96 to +1.96 constitute what is called the critical region or the region of significance. Alphas of .05 or less are critical values.

degrees of freedom. The number of independent comparisons available to estimate a specific parameter.

dependent variables. Measurements on the output of a process are called dependent variables.

design of experiments. The arrangement in which an experimental program is to be conducted, and the selection of the versions (levels) of one or more factors or factor combinations to be included in the experiment.

detection. The attempt to identify unacceptable output after it has been produced and then separating it from the good output.

dispersion. The amount of spread or variation of individual values around their average.

distributions. The standard of comparisons for statistical analysis such as normal, binomial, poisson, or chi square.

error of the first kind. Looking for an assignable cause when no such cause exists.

error of the second kind. Not looking for an assignable cause when such causes do exist.

experimental error. The residual component of a data set remaining after all the other components of the other data have been explained (caused by extraneous variables) other than those due to factors and blocks, which adds a degree of uncertainty to the observed response value.

F-ratio. The ratio of variances for pairs of samples taken from the same population. Used to determine if the populations from which the two samples were taken have the same standard deviation.

factor. An assignable cause that may affect the responses (dependent variables) and of which different versions (levels) are included in the experiment.

factorial experiment. An experiment in which the levels of each factor are combined with all levels of every other factor.

flow chart. A graphical representation of events and information that occurs in a series of actions or operations.

frequency distribution. A set of all the various values that individual observations may have and the frequency of their occurrence in the sample or population.

histogram. A plot of a frequency distribution in the form of rectangles whose bases are equal to the cell interval and whose areas are proportional to the frequencies.

hypothesis. An unproved proposition advanced as possibly true and consistent with known data but needing further investigation.

independent variables. Measurements on the input and process variables introduced into the data file.

interaction. In the analysis of variance, the tendency for the combination of factors to produce a result that is different from the mere sum of their two individual contributions.

kurtosis. A measure of the shape of a distribution.

log plot. The use of special graph paper for which one of the scales is logarithmic. Used with processes that are rate dependent or exponential in nature.

mean. The average value of some variable.

median. The middle value in a group of measurements when arranged from lowest to highest.

mode. The most frequent value of the variable.

moving range chart. A control chart for individual measurements with control limits based on the moving range.

nonconformities. Units that do not conform to specification or other inspection standards; sometimes called discrepant or defective units.

null hypothesis. The hypothesis, tested in a test of significance, that there is no difference (null) between the population of the sample and the specified population. The null hypothesis can never be proved true. It can, however, be shown, with specified risks of error, to be untrue.

observed value. The particular value of a characteristic determined as a result of a test or measurement.

p-chart. A plot of the proportion of the attributes counted to be the total number of units sampled.

parameter. A constant or coefficient that describes some characteristic of a population (e.g., standard deviation, average, regression coefficient.)

Pareto chart. An analysis of the frequency of occurrence that involves ranking all potential problem areas or sources of variation according to their contribution to cost or total variation. It is used to prioritize effort by concentrating on the *Vital Few* causes, temporarily ignoring the *Trivial Many*.

Poisson distribution. A distribution of events that results when the proportion of occurrence is small compared to the chance of occurrence.

population. The totality of items or units of material under consideration.

probability distribution. A distribution of relative frequencies.

probability plot. Data plotted on a special graph paper on which a normal distribution plots as a straight line.

process. A process is any set of conditions, or set of causes, that work together to produce a given result.

process capability. The limits within which a tool or process operates to turn out similar parts that can be maintained under control for a sustained period of time, under a set of given set of conditions. In turn, it is compared to a set of given specification limits (the basic use of control charts).

process control chart. A chart used to maintain a pre-determined distribution over a long period of time.

process flow chart. A breakdown of a process to its elements so that the details and sequence of operations can be examined with the aim of improvement or prearrangement.

quality. The totality of features and characteristics of a product or service to be satisfy a given need. What kind is it?

quality assurance. All those planned or systematic actions necessary to provide adequate confidence that a product or service will satisfy given needs.

quality control. A program that implies regular, systematic, and continuing application of control charts (and other closely related techniques and activities) to sustain a quality of product or service that will satisfy given needs.

random sample. Samples selected so as to minimize all sources of variation other than the factor being studied. This can be accomplished by having each sub-group consist of consecutive units as produced.

randomized block. A simple designed experiment that allows for all combinations of factors.

randomness. A condition in which individual values vary with no discernible or predictable pattern, although they may come from a definable distribution.

range. The difference between the largest value and the smallest value in a set of observations.

rational sub-group. Represents, for rational or logical reasons, a small sample that is free as possible from assignable causes representing a homogenous set of conditions.

regression. The tendency for the mean of given values for one variable to vary with others.

replication. Repeating an experiment to obtain a measure of precision.

residual variance. The remaining variance of a process after all known source variations are accounted for. It includes experimental error and assignable sources of variation not taken into account by the experiment.

response variable. A dependent variable.

sample. A group of units, portion of material, or observations that serves to provide information that may be used as a basis for making a decision concerning the larger quantity.

scatter plot. Shows the cause and effect relationship between one cause and the other.

sigma. When the upper case form (Σ) is used, it means to sum. The lower case form (σ) is used to designate a standard deviation.

significant. When a result deviates from some hypothetical value by more than can reasonably be attributed to the chance errors of sampling.

skewed. Lopsided, asymmetrical.

specifications. The required properties of a product to assure its function to meet the needs and expectations of the customer.

spread. The extent by which values in a distribution differ from one another: dispersion, range, variance.

standard deviation. A measure of the spread of a process output, the square root of the variance; denoted by the lower case Greek letter sigma σ for the population standard deviation and an by s for the sample standard deviation.

statistic. A quantity calculated from a sample of observations, most often used to form an estimate of some population parameter.

statistical control. To maintain a process within boundaries. A condition of a process from which all special causes of variation have been removed and only common causes remain. The absence of points beyond the control limits on a control chart and the absence of trends and non-random patterns within the control limits.

statistical process control. The use of statistical techniques such as probability plots, control charts, design of experiments, and analysis of variance to analyze a process so that appropriate actions may be taken to achieve and maintain a state of statistical control and to improve the process capability.

stem-and-leaf display. A special case of a frequency distribution that allows for the preservation of all the original sample data.

stratification. Systematic sampling done in such a way that two or more distributions are represented.

student's t. The ratio of the deviation of the mean of a sample of (n) individuals from an expected value to its standard deviation (σ/n). The t-distribution is expressed as a table for a given number of degrees of freedom (n - 1).

t-distribution. The normal distribution for sample data.

treatments. A combination of levels of each factor assigned to an experimental unit.

truncated distribution. A result when data has been cut short, such as the removal of data by prior inspection.

U chart. Measures the number of nonconformities (discrepancies or defects) that can have varying sample sizes.

universe. A group of populations from which samples are drawn, often reflecting different characteristics of the items or material under consideration.

variance. There are two types. Population variance (σ^2) is the measure of variability (dispersion) of observations based on the mean of the squared deviations from the arithmetic mean. Sample variance (s^2) is the measure of variability (dispersion) of observations in a sample based on the squared deviations from the arithmetic average divided by the degrees of freedom.

X-bar and R-chart. A control chart of the average and range of subgroups of data using variable measurements.

Z-distribution. Compares a random sample of one or more measurements with a large parent group whose mean and standard deviation are known.

Bibliography

Andrews, D. H. and R. H. Kokes. 1965. *Fundamental chemistry*. New York: John Wiley and Sons.

Arbor, Inc. 1987. Customer window grid. *Quality Progress*. June.

Fisher, Ronald A. 1932. The design of experiments. *Rothamsted experimental station 1932 annual report*. Indianapolis: Hafner Press.

Greenwood, J.A. 1950. Sample size for estimating the standard deviation as a % of its value. *Journal of the American Statistical Association*. V.45.

Goode, H. H. and R. E. Machol. 1957. *Systems engineering*. New York: McGraw Hill.

King, James R. 1981. *Probability charts for decision making*. Tamwerth, N.H.: Teams.

Hunter, J. Stuart. 1978. *Statistics for experimenters*. New York: John Wiley and Sons.

Monod, Jacques. 1978. Essay on the natural philosophy of modern biology. *Selected papers in molecular biology*. San Diego: Academic Press.

Nelson, Wayne. 1979. *How to analyze data with simple plots*. Milwaukee: ASQC.

Statistical process control handbook. 1956. A T & T Technologies. Indianapolis: Western Electric.

Tukey, John W. 1977. *Exploratory data analysis*. Ann Arbor, Mich.: UMI Press.

Wiener, Norbert. 1961. *Cybernetics*. Cambridge, Mass.: MIT Press.

Index

historical data 49
Hunter, J. Stuart 47
hypothesis 45, 46, 128, 130, 132, 134, 140, 149

I

inconclusive pattern 104
independent variable 6, 7, 48, 52, 77, 138, 139, 141, 145, 153, 154, 157
individual measurement 99, 100, 109
induction heat treating machine 97, 149, 181
inductive thinking 34
industrial experiment 17, 19, 48
industrial experimentation 153, 154, 155, 186
industrial experimenter 154
inspection system 1, 3
interview 33

L

Latin square 159, 160, 162
level of significance 119, 133, 137, 145, 177
line of best fit 89, 90, 93, 97, 138, 164, 179, 180

M

manufacturing 1, 4, 12, 13, 17, 24, 38, 41, 43, 48, 68, 122, 168, 174
manufacturing process 1, 4, 12, 41, 48, 168
manufacturing variability 12
mean 57, 63, 71, 72, 73, 74, 75, 76, 77, 80, 83, 86, 87, 89, 95, 97, 98, 111, 115, 116, 117, 118, 120, 124, 130, 131, 132, 133, 134, 135, 137, 173, 175, 176
mean square 173, 177
measurements 6, 7, 8, 9, 10, 71, 72, 76, 77, 97, 99, 100, 109, 115, 116, 118, 130, 131, 139, 153, 155, 174
median 63, 65, 72, 73, 86, 91, 131, 132, 133
minimum sample size 130
mixture 97, 103, 106
mode 63, 72, 73, 76, 77, 92, 131
Monod, Jacques 9
moving range 99, 100, 101
moving range chart 101
multiple correlation 145
multiple probability plots 147, 149
multiple regression analysis 15, 146
multivariate methods 145

N

natural pattern 102, 103
negative number 54, 56, 57
nitrogen factors 165
normal curve 71, 76, 77, 78, 116, 117, 119
normal distribution 8, 9, 10, 11, 12, 27, 57, 71, 77, 79, 80, 81, 83, 86, 88, 89, 93, 101, 111, 116, 119, 131
null hypothesis 67, 134, 136, 144

O

observation 8, 48, 57, 59, 60, 63, 65, 67, 72, 74, 76, 78, 81, 98, 129, 132, 140, 143, 145, 154, 157, 158, 162, 163, 173, 174
one-sample analysis 130, 131, 132
ordinate 39

About the Author

Robert F. Brewer has more than forty years of experience in statistical process control. He worked with Bonnie Small of Western Electric on the original *Statistical Process Control Handbook*. A senior member of IIE and ASQC, he is now with Industrial Designs and Engineering Associates in Lakeville, Pennsylvania.

About the Series

More and more people entering today's work force are being asked to perform multiple tasks that may include those once limited to traditional industrial engineers. The Engineers in Business Series is about understanding and applying a combination of business skills and industrial engineering principles. It was developed to meet the needs of students, new industrial engineers, and individuals without engineering degrees interested in learning about the different aspects of the industrial engineering profession.

About EMP

ENGINEERING & MANAGEMENT PRESS (EMP) is the book publishing division of the Institute of Industrial Engineers. EMP was founded in 1981 as Industrial Engineering and Management Press (IE&MP). In 1995, IE&MP was "reengineered" as ENGINEERING & MANAGEMENT PRESS by an entirely new staff.

As both IE&MP and EMP, the press has a history of publishing successful titles. Recent successes include: *Toyota Production System, 2nd Edition*; *Beyond the Basics of Reengineering*; *Business Process Reengineering: Current Issues and Applications*; and *Managing Quality in America's Most Admired Companies*.

EMP's newest titles are *By What Method?*; *Simulation Made Easy*; and *Essential Career Skills for Engineers*, the first in the **Engineers In Business** Series, of which this book is a part.

For more information about EMP or to request a free catalog of EMP's current titles, please call IIE Member & Customer Service at 800-494-0460 or 770-449-0460.

About IIE

Founded in 1948, the Institute of Industrial Engineers (IIE) is comprised of more than 25,000 members throughout the United States and 89 other countries. IIE is the only international, nonprofit professional society dedicated to advancing the technical and managerial excellence of industrial engineers and all individuals involved in improving overall quality and productivity. IIE is committed to providing timely information about the profession to its membership, to professionals who practice industrial engineering skills, and to the general public.

IIE provides continuing education opportunities to members to keep them current on the latest technologies and systems that contribute to career advancement. The Institute provides products and services to aid in this endeavor, including professional magazines, periodicals, books, conferences and seminars. IIE is constantly working to be the best available resource for information about the industrial engineering profession.

For more information about membership in IIE, please contact IIE Member and Customer Service at 800-494-0460 or 770-449-0460 or http://cs@www.iienet.org.